Electrostatics Handbook

By

Charles Green

HOWARD W. SAMS & CO., INC.
THE BOBBS-MERRILL CO., INC.
INDIANAPOLIS · KANSAS CITY · NEW YORK

FIRST EDITION

FIRST PRINTING—1973

International Standard Book Number: 0-672-21023-1
Library of Congress Catalog Card Number: 73-88015

Preface

The study of electric charges at rest, under conditions where the positive and negative charges are separated from each other, is called "Electrostatics." A stationary charge is often called "static electricity," distinct from moving or "current" electricity which we use to power our electric lights, tv sets, and other electrical appliances in the home.

Electrostatic electricity was the only kind of electricity known for hundreds of years. Before "current" electricity became the servant of man, static electricity was a plaything of royal courts in Europe, as well as the object of study by scientists. Electrostatics has been considered to be only an era in the history of electricity, including the initial observation of the power of amber to attract lightweight substances and the invention of the lightning rod by Benjamin Franklin. Discoveries, such as the "Leyden jar," have been the basis for the capacitors used in radios and tv sets. Other basic discoveries, made during centuries of electrostatic study, have been the foundation for many of our modern-day industrial machines and home appliances.

This book is written for individuals who wish to learn the fundamentals of electrostatics by building demonstration projects. Each chapter includes groups of projects that cover different aspects of electrostatic electricity generation, detection, and use. The projects are designed for easy construction with a minimum of specialized tools and materials. They are suitable as ideas for science fair projects and for home experimentation.

CHARLES GREEN

To Clare, Susan, and David

Contents

CHAPTER 1

INTRODUCTION 7
History of Electrostatics—Theory—Electrostatic Induction—Electron Sink—Distribution of Electrical Charges—Experiments

CHAPTER 2

ELECTROSTATIC GENERATORS 23
Rotating Friction Generator—Electrophorus—Varley Machine—Wimshurst Electrostatic Generator—Kelvin Water-Drop Generator—Van de Graaff Generator

CHAPTER 3

ELECTROSTATIC DETECTORS 53
Pith-Ball Indicating Devices—Versorium—Gold-Leaf Electroscope—Shielded Electroscope—Twin-Electrode Foil Electroscope—Electrostatic Voltmeter

CHAPTER 4

FROM LEYDEN JARS TO CAPACITORS 73
Water Bottle Leyden Jar—Leyden Jar—Franklin's Leyden Jar Dielectric Experiment—Paper Capacitor Experiment

CHAPTER 5

ELECTROSTATIC MOTORS 83
Franklin Motor—Plastic-Disc Motor—Plastic-Drum Motor—Two-Brush Disc Motor—Four-Brush Disc Motor

CHAPTER 6

ELECTROSTATICS AND RADIATION 101
Early Forms of Radiation Detection—Electroscope Test for Radioactivity—Cloud Chamber—Types of Vapor Trails

CHAPTER 7

MODERN ELECTROSTATIC DEVICES 109
Possible Future Uses of Electrostatics—Simulated Lightning Rod Experiment—Smoke Removal Precipitator—Electrostatic Speaker

APPENDIX

CONSTRUCTION NOTES 123
Metal Knobs—Plastics

INDEX 125

CHAPTER 1

Introduction

Electrostatic electricity has always been one of the great natural forces. Everyone is familiar with the awesome lightning bolts of a summer electrical storm and the related effects of static electricity on a much smaller scale. Static electricity causes the slight electrical shock received after walking across a carpet and touching a metal door knob (Fig. 1-1). The same type of shock will result if you slide across a plastic-covered seat in an automobile and then touch the door handle.

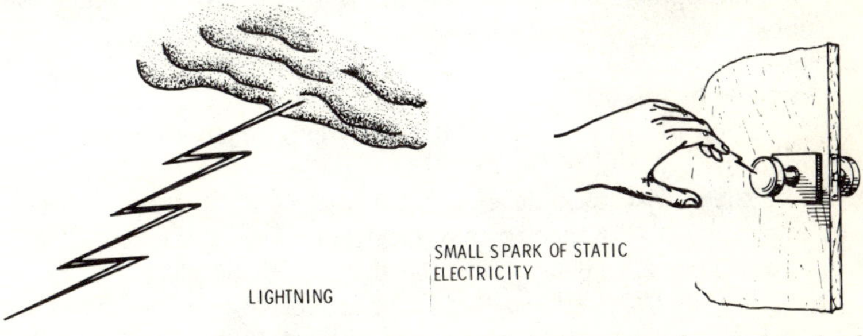

Fig. 1-1. Lightning and static electrical spark between hand and door knob are forms of electrostatic energy.

We live in the midst of seen and unseen electrical forces. The atmosphere acts as a gigantic generator of electrostatic electricity (similar to the Van de Graaff generator project in Chapter 2). Its voltage, in respect to the surface of the earth, increases many thousands of volts, until it reaches a potential of hundreds of thousands of volts at the ionosphere. On hot, low-humidity days, electrical shocks can be received from radio antennas because of the atmospheric electrostatic-potential buildup on the antenna wires.

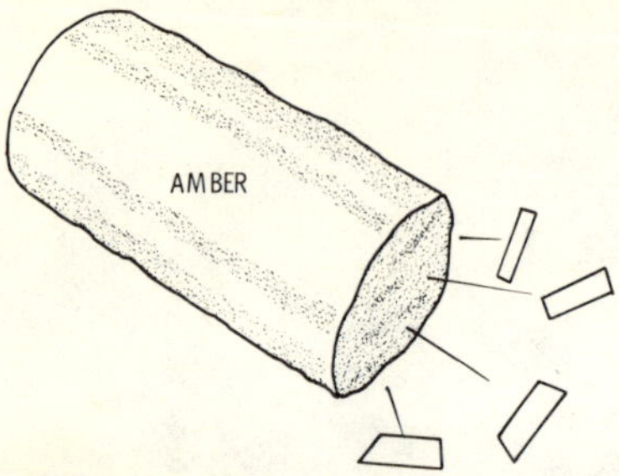

Fig. 1-2. Attraction of rubbed amber for lightweight objects was discovered by the ancient Greeks.

Electrostatic electricity operates according to the same fixed principles, or laws, whether the electricity is in the form of lightning or the attraction of lightweight objects. The early Greek philosophers observed that rubbed amber would attract lightweight objects, such as lint, feathers, and paper (Fig. 1-2).

HISTORY OF ELECTROSTATICS

The ancient Greek philosophers, by sheer intuition, made remarkable deductions; our civilization has only in the last few hundred years proven them to be facts. For example, 2500 years ago, the Greek philosophers Leucippus and Democratus presented a theory that all matter was made up of very tiny particles or atoms. In 600 B.C., the Greek philosopher Thales experimented with the attraction that a piece of rubber amber had on small pieces of feathers and straw.

The atomic theory, a piece of fossilized resin (amber), and the lightning bolts of electrical storms all intertwine in the history of electrostatics and in the search that early experimenters made for the common background to all these.

From the early Greeks to Queen Elizabeth's time, little is known of any experiments with electrostatic electricity. Perhaps medieval magicians investigated, and used, effects of electrostatic electricity as part of their bag of tricks. If so, they kept any discoveries to themselves. Queen Elizabeth's physician, William Gilbert, experimented in 1600 with rubbed amber and found other substances that had the ability to attract lightweight objects. He called these attractions "electric" after the Greek word for amber. Gilbert invented the first electrical indicat-

(*A*) *Gilbert's versorium.*

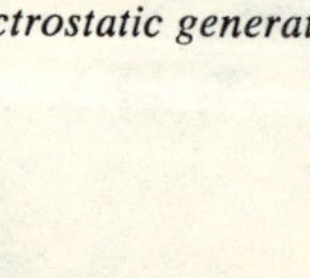

(*B*) *von Guericke's electrostatic generator.*

Fig. 1-3. The beginning of scientific investigation of electricity.

ing instrument. It was a lightweight, pivoted metal bar (Fig. 1-3A) that would be attracted toward a charged object that was "electrified" by rubbing.

In approximately 1660, Otto von Guericke of Magdeburg, Germany, invented the first frictional electrostatic generator (Fig. 1-3B). This generator (or electrical machine, as the early experimenters called generators) was composed of a rotating ball of sulfur that was rubbed to give the sulfur an electrical charge. The ball of sulfur was then lifted off the machine and used for his experiments.

In 1709, an English experimenter, Francis Hauksbee, built a more efficient electrostatic generator by using a glass ball in place of the sulfur ball. His was the first machine in which electrical sparks could be obtained. His experiments with a spinning, evacuated glass globe produced the first display of electrical light (similar to present day neon signs) by ionizing the residual gases with the high-voltage static electricity.

Other experimenters improved the electrostatic generator and provided leather cushions (in place of dry hands) for rubbing the rotating glass balls or cylinders. The electric charges were collected by hanging metal chains resting on the rotating glass ball.

In 1729, another Englishman, Stephen Gray, found that electrical charges can be transferred by contact and through "conductors" made from hempen cords. He made up a list of electrical conductors and insulators. Hempen cord is a poor conductor and probably worked as an electrical conductor because of the high voltages involved and the dampness of the climate in England.

Important experiments were made in 1733 by two Frenchmen, Abbe Nollet and Charles Francois de Cisternay Dufay (superintendent of the royal gardens of King Louis XV). It was discovered that there were two different types of electric charges—one kind emitted by rubbed glass, and the other type emitted by rubbed resin. It was found that two objects charged with the same type of electricity repelled each other, and that two objects charged with different types of electricity attracted each other.

In 1745, European scientists made independent experiments dealing with the electrical charging of a glass flask filled with water. Among these scientists were E. G. von Kleist, dean of the cathedral at Kummin, Pomerania, and Professor van Musschenbroek of the University of Leiden, Holland, for which the Leyden jar was named (Fig. 1-4A).

In 1747, the great American scientist, Benjamin Franklin, discovered that the electrical charge is stored in the glass walls of the Leyden jar (not in the inner or outer conducting materials). This discovery forms the basic theory of the modern-day capacitor. He designated electrical charges from rubber glass as positive (+), and electrical charges from rubbed resin (similar to modern-day rubbed plastics in nature) as negative (−). He also invented the lightning rod (Fig. 1-4B) and the first electric motor.

Following Benjamin Franklin's discoveries, Franz Ulrich Theodor Aepinus, a professor at the Russian Imperial Academy in St. Petersburg, discovered that two parallel metal plates separated by an air gap stored electrical charges in the same manner as a Leyden jar. (The air acted as the dielectric in place of the glass of the Leyden jar.) This was the forerunner of the modern-day variable capacitors used in radios.

Fig. 1-4. Basic electrical discoveries.

(*A*) *von Musschenbroek's Leyden jar.* (*B*) *Franklin's lightning rod.*

Fig. 1-4. Basic electrical discoveries.

In 1775, Alessandro Volta invented the first induction type of electrostatic generator, the "electrophorus." This type of generator was also called an "influence machine" by early experimenters. Volta invented the first chemical electric battery, in 1800, and gave his name to the unit of electromotive force, the volt.

Charles Augustin de Coulomb, a French physicist, researched electric charges; in 1785, he found that the attraction (or repulsion) between small charged spheres (or points) is inversely proportional to the square of their distance. The coulomb, a unit of quantity (or amount) of electricity, is named after him.

Abraham Bennet, an English experimenter, invented the gold-leaf electroscope in 1787. It has been used in various versions, up to modern times, as the main instrument for the investigation of electric charges.

In 1820, Hans Christian Oerstead, Danish physicist, discovered that a current of electricity creates a magnetic field. André Marie Ampère, a French scientist, investigated the nature of electric current (the unit of electric current, the ampere, is named after him). In 1827, Georg Simon Ohm, a German physicist, discovered the basic relationship between electrical current, voltage, and resistance (Ohm's law).

Michael Faraday, English physicist, is famous for his discovery in 1832 of electromagnetic induction (independently of the American, Joseph Henry). Faraday also made important experiments in electrostatic electricity. He introduced the idea of lines of force in the electric field around charged bodies.

A rotating inductive electrostatic generator was invented by a British physicist, C. F. Varley, in 1860. The generator used a rotating glass disc which, in principle, may be considered to be similar to a rotating electrophorus. A much more elaborate and more efficient inductive generator was invented by James Wimshurst, an English engineer. His generator used two glass discs, rotating in opposite directions and generating very powerful sparks. This type of generator was used, until modern times, to produce high voltages for X-ray machines. This generator was supplanted, in 1929, by the belt-driven generator of the American physicist, Robert J. Van de Graaff.

During the 19th century, British physicist Lord Kelvin (William Thomson) did research into the nature of electricity. He invented the mirror astatic galvanometer in 1858 and an electrostatic voltmeter in 1867. He also invented the "water-drop electrostatic generator."

In the early part of the 20th century, British scientist Lord Rutherford and American physicist Robert J. Millikan, among others, did research which developed the modern theories about the atom and the nature of electrical charges.

THEORY

In any substance, the smallest particle that can exist by itself and still have all the characteristics of that substance is called a "molecule." For example, if a drop of water is divided and divided, the last possible division that could result in a product of water would be called a water molecule.

Molecules are composed of smaller particles called atoms. Atoms can be defined as the smallest part of an element that can take part in ordinary chemical changes. A molecule of water is composed of two atoms of hydrogen and one atom of oxygen (Fig. 1-5).

Atoms are composed of still smaller units called protons, electrons, and neutrons. The proton has a basic positive (+) electrical charge, and the electron has a basic negative (−) electrical charge. The neutron has no electrical charge and is electrically neutral.

All atoms are made up of different amounts of these smaller units. The simplest atom, that of hydrogen, is composed of one proton, and one electron revolving around the proton (which acts as a nucleus) in an orbit or "shell." Because the positive charge of the electron, the atom is electrically balanced, or neutral (Fig. 1-6).

Normally, an atom is electrically balanced—the proton positive charge and electron negative charge equal each other. However, some metal atoms (copper, silver, and aluminum) have one or more electrons that are loosely held in the outer shell. These electrons are called "free electrons," and are easily shifted from one atom to the other. The protons are much heavier than the electrons, and they are relatively

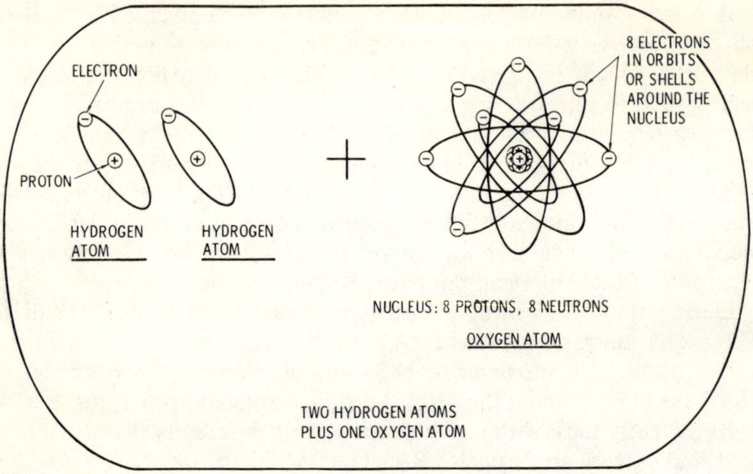

Fig. 1-5. Water molecule.

immobile. The concerted movments of numbers of free electrons make up what is known as an electric current. The metals with free electrons are called "conductors" (Fig. 1-7).

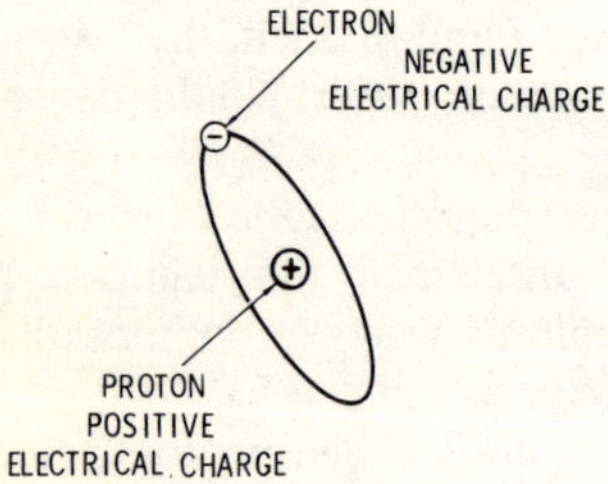

Fig. 1-6. Hydrogen atom.

When electrons are fed into one end of a conductor, such as a copper wire, they will force the closest free electrons in the outer shell of an atom to shift position, and they in turn will force other free electrons to move from nearby atoms. This action of electric current flow will continue along the complete length of the copper wire at a relative velocity near the speed of light. The current flow via the conductor will be from a source of an excess of electrons, which is negative, to

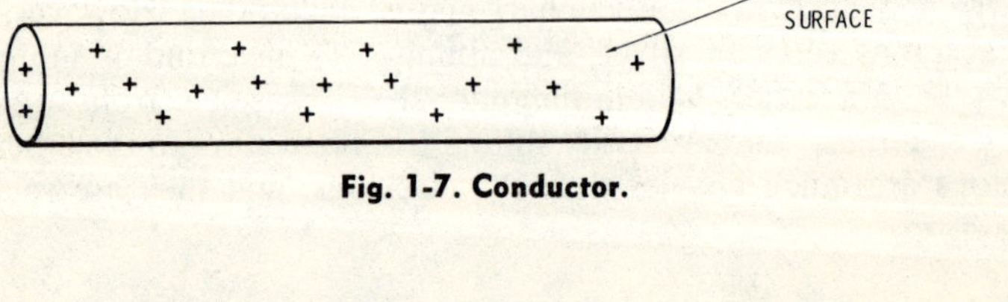

Fig. 1-7. Conductor.

that of a deficiency of electrons, which is positive. The pressure that forces electrons to move is called "potential," and is expressed as "voltage." Electrostatics is normally high voltage (high potential), with intermittent or low current flow.

A conductor can also have a specific negative or positive electrical charge. This is accomplished either by taking electrons away, leaving the conductor with a positive charge, or by adding electrons to the conductor, giving it a negative charge. The conductor cannot have both charges at the same time; they will cancel each other, and the conductor will then become neutral.

Ions

An ion is an electrically charged atom. Positively charged ions are atoms that have lost some free electrons, and negatively charged ions are atoms that have gained some free electrons.

Ions can be formed in gases by electrical sparks, a strong electrostatic field, or collisions between particles, such as electrons. The term "ionization" means the formation of ions.

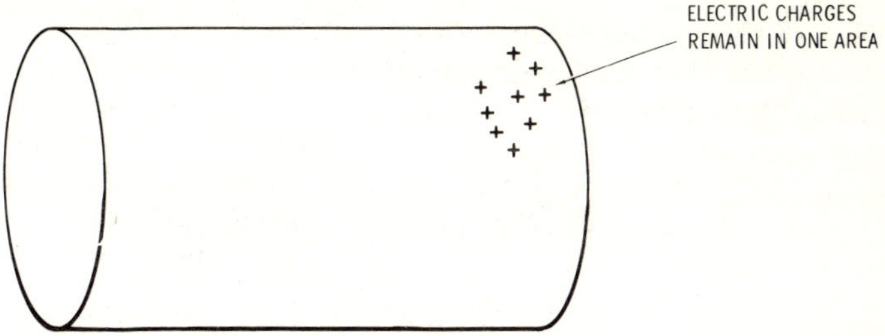

Fig. 1-8. Insulator.

If the air (or gas) is ionized to a great extent by an electrostatic field or discharge, the air will become luminous around the conductor of the electrostatic field or discharge. This effect is also called a "corona discharge." If the ionization of the air becomes greater than the point of corona discharge, the air becomes conductive and a spark will discharge to ground or another conductor of lower electrostatic potential.

Glow Lamps

The effect of ionization of gases to create a luminosity is utilized in "glow tubes," such as neon lamps. These lamps usually have two metal electrodes and a small amount of neon gas in a glass bulb. The neon lamps (particularly the small NE-2 type) are very useful to indicate the presence of electrostatic fields, or potential. They are used in many projects in this book.

Materials having electrons that are not easily moved and those with few free electrons are called "insulators" (Fig. 1-8). Glass, ceramic, silk, wool, and plastic are good examples of these electrical insulators. However, these insulators can be made to release electrons by mechanical force, such as by rubbing with a suitable material; they can also have electrons mechanically transferred to the surface of the insulator. These electrons will remain in the area to which they were transferred and will not spread over the surface of the insulator as they do in a conductor. Insulators are also called "dielectrics."

Electrically charged objects have theoretical electric lines of force about them, just as a magnet has a magnetic field with magnetic lines of force about it. The greater the electrical charge, the larger and stronger the electric field (Fig. 1-9).

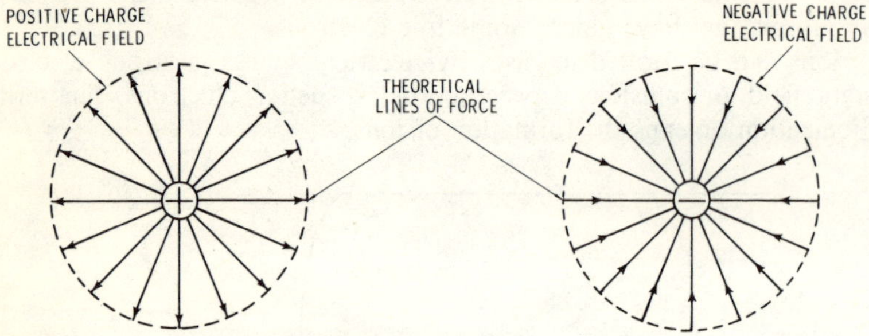

Fig. 1-9. Electric fields about charged objects.

The normal state of most material is electrically neutral (no electrical charge). If the electrical balance of a material is disturbed by friction with another material and causes a movement of electrons between the materials, the resulting electrical charges on the materials will be equal, and opposite, in electrical polarity.

For example, if a plastic rod is rubbed with a woolen cloth, electrons from the cloth will move to the plastic rod. Therefore, the cloth will have a positive charge (absence of electrons), and the plastic rod will have an equivalent negative charge (more electrons), as shown in Fig. 1-10.

The cloth will soon lose its electrical charge because it is a poor insulator, and it will probably have absorbed some moisture from your hand, making it an electrical conductor. The charge will then be dissipated by electrons moving from your hand to the cloth, making the cloth electrically neutral again. The rod will hold its electrical charge for some time because plastic is a good insulator.

The negative electrical charge on the plastic rod (or any charged item) can be transferred to a neutral object by touching the object with the rod. Then electrons from the rod will flow to the neutral object,

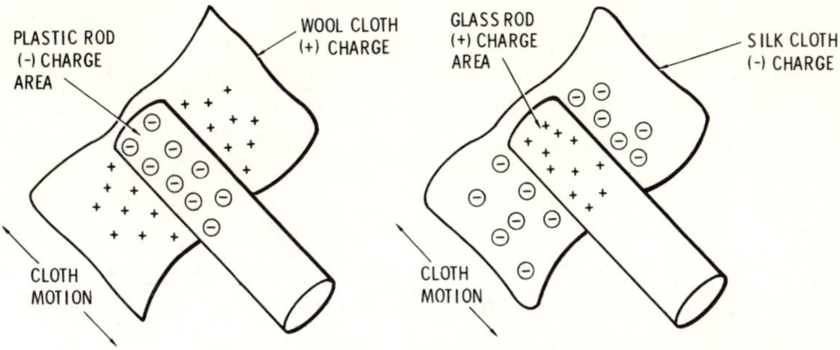

Fig. 1-10. Electrifying by rubbing (friction).

causing the neutral object to have a negative charge. This is known as "electrostatic charging by contact" (Fig. 1-11). If the charged item is positive, electrons from the neutral object will flow to the rod, leaving the neutral object with a positive charge.

Fig. 1-11. Charging by contact.

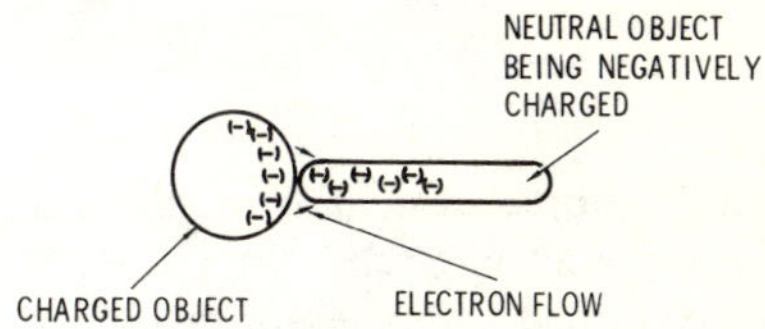

ELECTROSTATIC INDUCTION

Electrically charged objects have an electrostatic field surrounding them. If two charged objects have opposite polarities, their field will merge (Fig. 1-12), and the objects will try to move together (mutual attraction, Fig. 1-13). If the charges on the objects are the same polarity, the electrostatic fields will curve away from each other (Fig. 1-14), and the objects will try to move away from each other (mutual repulsion, Fig. 1-15).

An electrostatic field surrounding a positively charged object (positive electrostatic field) will cause electrons, on a neutral body within

Fig. 1-12. Electric fields about two nearby charged objects with unlike polarities.

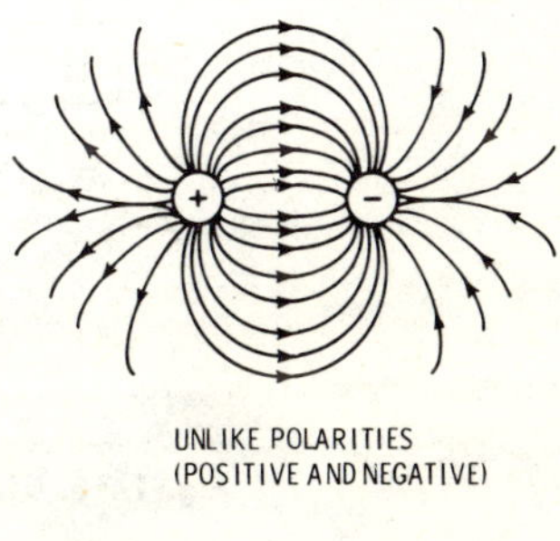

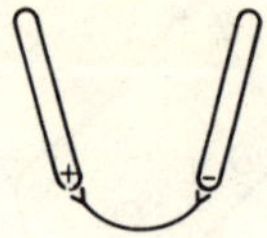

Fig. 1-13. Unlike charges attract.

the positive electric field, to be attracted toward the positively charged object. If the distance is too great for the electrons to move over any intervening space (or by contact), the electrons will accumulate on the side of the neutral body closest to the positively charged object.

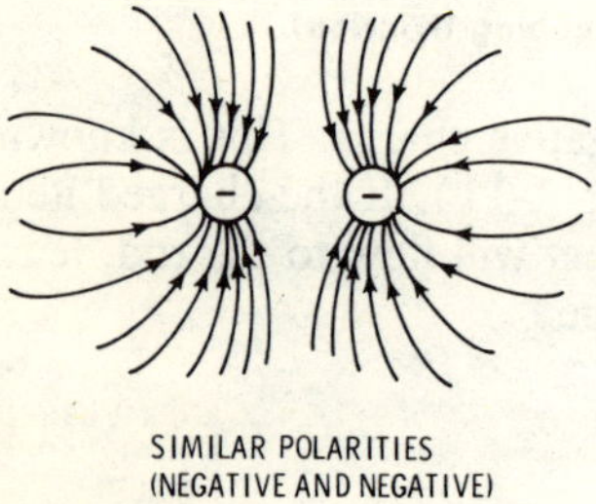

Fig. 1-14. Electric fields about nearby charged objects with similar polarities.

If the object is negatively charged, the electrons on the neutral body will move away to the other side, the side farthest from the negatively charged object. If a temporary connection is made to ground from the

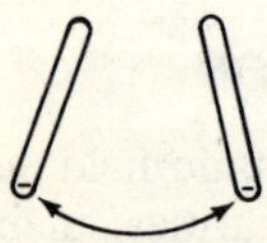

Fig. 1-15. Like charges repel.

side of the neutral body farthest away from the charged object, the electrons on that side will flow to ground, leaving the former neutral body with a positive electrical charge (Fig. 1-16). In a positive electro-

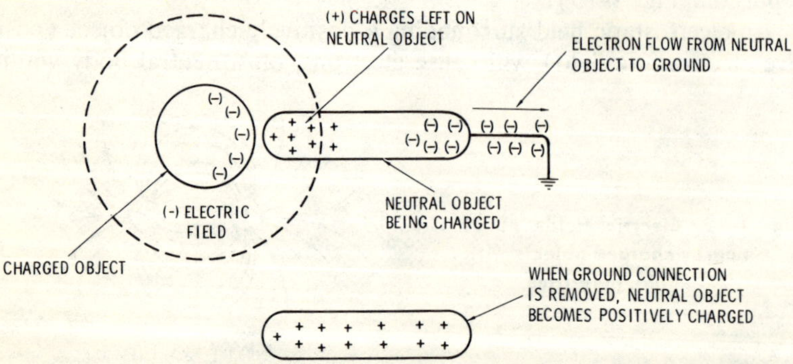

Fig. 1-16. Charging by induction.

static field, the electrons will flow from ground to the body, giving the body a negative charge.

In summary, there are three methods of electrostatic excitation (charging or electrifying):

Friction. A static charge is best obtained by rubbing a hard, nonconducting material (electrical insulator) with a soft nonconductor. Some materials that seem to work best together are glass rubbed with silk (glass assumes a positive charge) and plastic rubbed with wool (plastic assumes a negative charge).

Contact. The neutral body in contact with the charged object assumes the same polarity.

Induction. The neutral body assumes the opposite polarity as the charged object.

ELECTRON SINK

The earth is considered to be electrically neutral because of the enormous amount of electrons that would be needed to change its electrical charge significantly. Electrons can flow to the earth or flow from it to any electrostatically charged object without changing the earth from its neutral state. Objects connected to the earth may assume a temporary electrical charge (for example, during an electrical storm). For all practical purposes, the earth may be considered to be an "electron sink," as it absorbs electrons (electrical ground).

Your body can also act as an electron sink to a limited degree. Because of the limited amount of flowing electrons, you can absorb small electrostatic charges without harm. But at times the electrostatic charges can accumulate to a great degree and cause a severe shock. Caution is advised in electrostatic experiments involving large charges.

At the high voltages and low currents involved in electrostatics, wood and any moist material (such as moisture film on glass) can act as a conductor. So, electrically charged objects will gradually lose their charge through the tabletop, floor, cement, and other household building materials to the earth electrical ground.

DISTRIBUTION OF ELECTRICAL CHARGES

When an electrical insulator has a charge, the charge stays in the same area in which it was first obtained (contact, friction, or induction). But, a charge on an electrical conductor spreads over the surface (not inside). If the conductor is a smooth round ball, the electric charge will be spread evenly over the surface. If the conductor has sharp points, great curvature, or irregularities, the charge will accumulate in these areas. If the electrical charge is high enough, the charge will even spray off the sharp points of a conductor. The more sharply pointed a conductor is, the faster it will discharge. A rounded surface will hold its charge much longer than an irregular or pointed surface; a larger rounded surface will hold its charge longer than a smaller rounded surface.

EXPERIMENTS

Electrifying a Plastic Comb

Cut a sheet of paper into small, about ¼-inch square, pieces and pile them on a wood table. Hold one end of a plastic comb (or a section of any plastic item) and rub the other end with a woolen cloth (Fig. 1-17).

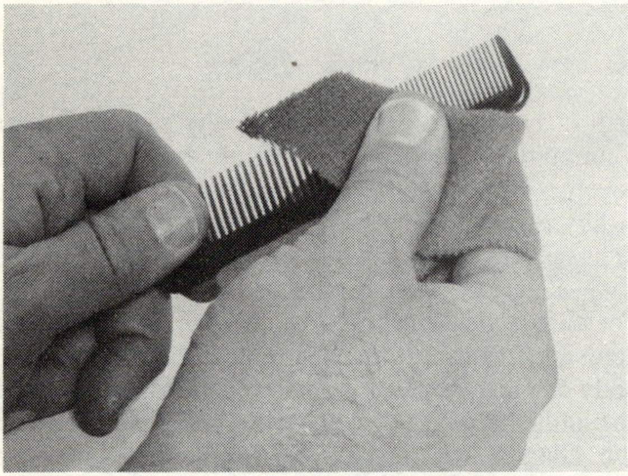

Fig. 1-17. Electrifying a plastic comb by rubbing it with a woolen cloth.

Move the rubbed (electrified) end of the comb near the pile of paper and observe what happens; you should see that some of the pieces of paper are attracted to the comb (Fig. 1-18). Note: if nothing happens, make sure that the paper, comb, and cloth are perfectly dry. If necessary, try heating the cloth to eliminate any moisture.

If the electrical charge on the comb is strong enough, some of the bits of paper will be attracted to the comb and then fly off. This is due to some of the paper bits acquiring the same charge polarity as the comb. Then, as like charges repel, the paper bits are repelled by the comb.

You can see very dramatic results by substituting salt in place of the paper. The salt grains, being small and lightweight, will easily move and bounce around. Try this experiment with other household items such as sugar, dry food grains, bits of thread, etc. The early experimenters found that many items could be electrified. Also, try the experiment with a glass rod and a silk or nylon cloth. The electrical charges will be different; most types of plastic have a negative charge and glass has a positive charge. However, the experiments should work in the same way, except that the glass may have to be rubbed longer; it may be harder to electrify than plastic.

Conduction Experiment

Place a metal table knife on top of a dry drinking glass and cut a length of paper approximately ⅛-inch wide and 5 inches long. Hold the strip of paper near the end of the knife and touch the other end of the knife with a comb that has been electrified by rubbing with a cloth. Observe that the strip of paper is attracted toward the knife when the comb touches.

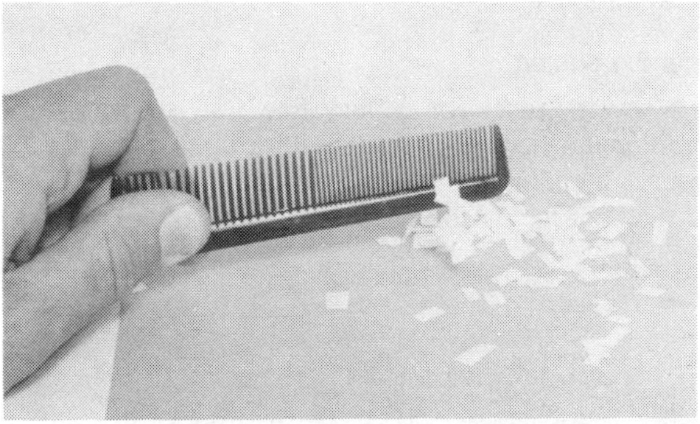

Fig. 1-18. Attracting small sections of paper with an electrostatically charged plastic comb.

This experiment show that a metal object can conduct the electrostatic charge. The paper strip becomes attracted to the electrostatic charge of the comb just as it would if it was placed close to the comb. If the comb charge is strong enough, you can still attract the paper to the knife, even after the comb is removed from the end of the knife (Fig. 1-19). This is because some of the electrical charge remained in the knife and was not conducted away to ground. The drinking glass acted as an electrical insulator and the electrical charge could not flow through it.

If the paper strip makes contact with the knife, the paper will become charged with the same polarity and be repelled away from the knife. You can repeat the experiment with an electrified glass rod and observe that the metal conducts the positive charge of the glass rod as well as it did the negative charge of the plastic.

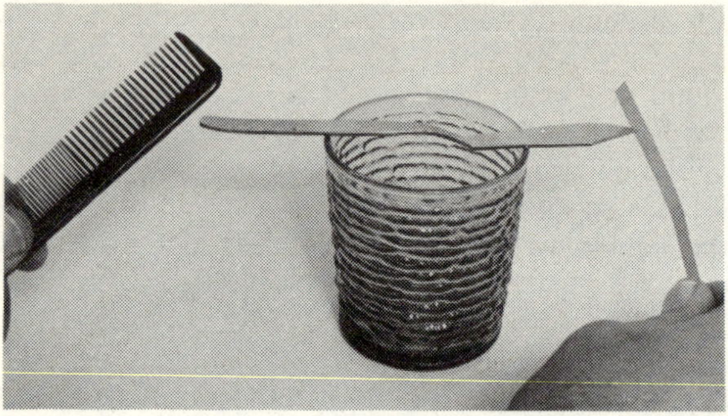

Fig. 1-19. Performing the conduction experiment with a plastic comb.

Note that glass or plastic will act as an electrical insulator (no charges will flow through), unless it is electrified by rubbing or other means of excitation.

Water-Drop Experiment

Adjust a water faucet for a very fine flow of water and bring a plastic comb, electrified by rubbing with a cloth, near the water. You should observe a change in the direction of water flow near the comb. If the electrical charge on the comb is strong enough, the water stream

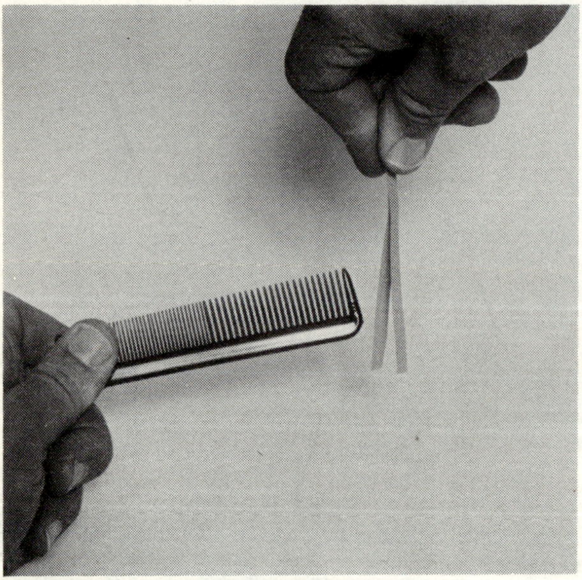

Fig. 1-20. Repulsion of like charges with an electrified plastic comb.

will actually bend sharply. This is because of the attraction of the electrons in the water molecules toward the positive charge of the comb.

You will also notice small water drops on the surface of the comb; the water drops will act as small charge carriers and move accordingly. (See the Kelvin water-drop generator project in Chapter 2). Try the experiment with an electrified glass rod.

Repulsion of Like Charges Experiment

Cut two strips of paper approximately ⅛-inch wide and about 4 inches long. Set the paper aside temporarily and electrify a plastic comb by rubbing it with a cloth.

Hold the top of the strips together with one hand and touch the bottom of the strips with the electrified comb. Observe that the strips move apart because they each have the same polarity of charge from the comb (Fig. 1-20). After a while the strips will lose their charge (it will leak off the paper to your hand) and will move together to approximately their former position.

You can try the same experiment with two strips of plastic. Charge them by rubbing them with a cloth or by touching them with an electrified comb. The plastic strips should hold their electric charges longer than the paper strips, as plastic is a better insulator than paper.

CHAPTER 2

Electrostatic Generators

The invention of the first electrostatic generator by Otto von Guericke, a 17th-century mayor of Magdeburg, Germany, ranks as one of the cornerstones of modern-day technology. The question is, why did this invention have to wait so long? The process was quite simple. A glass globe was filled with molten sulfur and then broken, after the sulfur cooled, to create a sulfur globe. The sulfur globe was then perforated so that it could spin on an iron axle. This entire operation could have been done in ancient times.

The electrostatic energy developed by this crude generator was much more than could be created by the drudgery of rubbing amber, and the scientists of the day could perform many more experiments. The basic invention by von Guericke was improved by many experimenters, such as Francis Hauksbee.

In this chapter, you can follow the evolution of the electrostatic generator from the simple friction type, similar to the one used by Francis Hauksbee, to Volta's electrophorus, Varley's rotating induction generator, Kelvin's water-drop generator, and the modern Van de Graaff generator.

You can use the generator projects (Fig. 2-1) in this chapter, particularly the Van de Graaff generator, as electrostatic power sources for experiments in the following chapters of this book.

ROTATING FRICTION GENERATOR

Construction of the Rotating Friction Generator

Instead of the sulfur ball used by Otto Von Guericke or the glass globe used by Ben Franklin and other early experimenters in their electrostatic generators, we used a plastic jar (Fig. 2-2). It is easier

Fig. 2-1. Some of the electrostatic generator projects included in this chapter: (Left to right) Van de Graaff generator, Varley machine, rotating friction generator.

to drill holes in and the plastic jar will normally require less rubbing than glass does to acquire an electric charge.

Begin construction by selecting a suitable plastic jar with a tightly fitting screw-on plastic lid. Since some plastics will generate electro-

Fig. 2-2. Rotating friction generator project.

static electricity better than others, test various jars by rubbing them with a woolen cloth and select the one that shows the greatest attraction for small pieces of paper. We used a plastic jar approximately 2¾ inches high and 2¾ inches in diameter, but any size will work; scale the construction to fit your size of jar. Drill centered holes in the bottom of the jar and in the lid to fit threaded rod I, as shown in the drawing of Fig. 2-3.

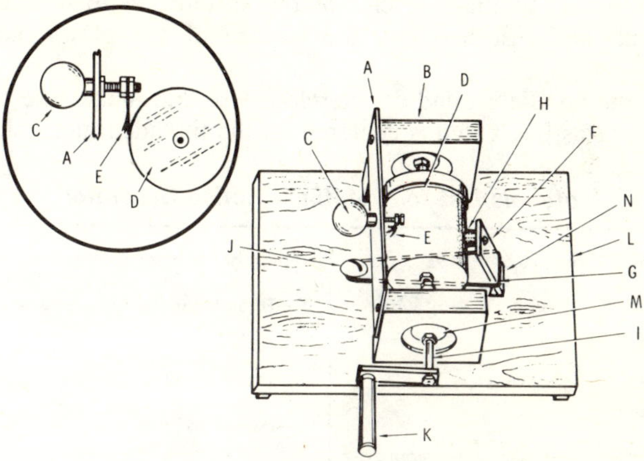

Fig. 2-3. Plans for the rotary friction generator.

Select a baseboard size to fit your size of jar; any type of wood can be used. For ease of construction, use a soft wood (such as white pine or particle board) and follow the general layout of our model. Cut the baseboard and B wood sections to size. Cut holes in the B sections to fit the M bearings. We used replacement roller skate wheels (available at many hardware stores) for the bearings. If no bearings are available, you can use short lengths of metal tubing that will fit loosely over the threaded rod (apply grease sparingly). Use set screws or cement in the B sections to hold the bearings in place.

Our model has crank handle K mounted with a short metal section for a counterclockwise rotation, but the crank handle can be installed on the opposite side for clockwise rotation. Assemble the jar, bearings, and handle on the threaded rod with nuts and washers outside the jar. If necessary, you can use a wood cylinder inside the jar to strengthen it to make mounting easier. Also, use nuts and washers on both sides of the bearings to keep the assembly from wobbling from side to side. Space the jar ends about 1 inch from each of the B sections. Make sure that the jar is centered between the two B sections before cementing or screwing the sections in place.

Use small, metal hinge N at the base of plywood section F and position it so that the end of F rests on the center of the plastic jar. Screw in rubber bumper H on the end of F so that it will ride along the surface of the plastic jar. Install wood screw J, in approximately the position shown in the drawing, to hold the end of rubber band G fitted around the base of F. Use only a light rubber band to avoid placing too much pressure on the surface of the plastic jar.

Cut plastic section A and install it across the top front of the two B sections as shown. Mount metal knob C and the solder lug with stranded wire E so that the end of the stranded wire is unraveled, spread out, and just touching the surface of the plastic jar as it revolves.

Crank the handle around counterclockwise and make sure that the plastic jar turns freely and is centered on the threaded metal shaft.

Materials List for Rotating Friction Generator

A—1/8" plastic section 6" × 1 1/2"	H—Rubber bumper to fit F (See text.)
B—3/4" wood section 3" × 2 1/2" with center hole to fit M	I —Threaded metal rod approximately 8" long to fit M
C—Metal knob approximately 1 1/4" diameter	J —Wood screw to hold the end of G
D—Plastic jar with plastic lid approximately 1 1/4" diameter	K—Handle to fit I
E—Solder lug with 1/2" stranded wire	L —3/4" wood base 7 1/2" × 10"
F—1/4" plywood section 3" × 1 1/2"	M—Ball bearing roller skate wheels (See text.)
G—Rubber band	N—Metal hinge to fit F

Adjustment and Test of the Rotating Friction Generator

Tape (or fasten with a rubber band) a section of woolen cloth around the rubber bumper, mounted on the end of the hinged F wood section. The cloth should be between the rubber and the plastic jar surface. Adjust the position of the rubber band on F to adjust the pressure of the woolen cloth on the jar. Allow for just enough pressure to cause fast rotation of the jar. You may have to try several rubber bands for the right tension.

Rotate the generator so that the jar revolves (counterclockwise in the model) from the woolen cloth toward the wire brush E. Adjust the wire brush ends so that they are curved in the direction of the jar rotation and so that the wire ends are as close as possible to the jar, without touching the jar surface. Fan the wire ends out away from each other.

Rotate the generator rapidly several times and then hold an electroscope (see Chapter 3) near metal ball C. The electroscope leaves

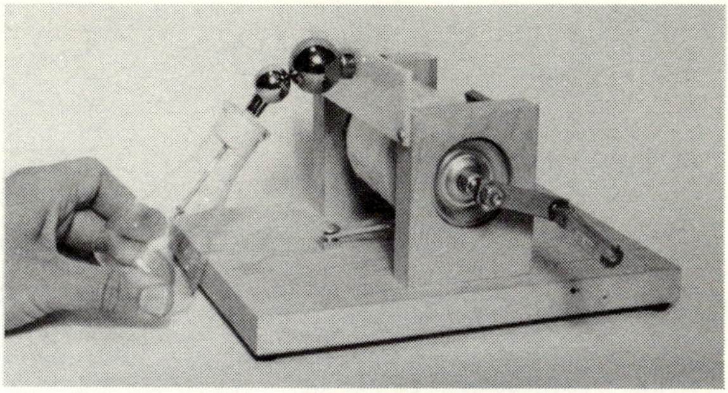

Fig. 2-4. Testing the rotating friction generator with an electroscope.

should move apart, indicating that the generator is functioning (Fig. 2-4). Try connecting one lead of a NE-2 neon lamp to the metal ball while holding the other lead. The neon should flash momentarily.

Try using other types of cloth and check electroscope deflection and neon lamp operation. Also, you can experiment by taping other materials (such as plastic household wrapping) over the jar surface and noting results. Always make sure that all materials are clean and dry.

Theory

As the handle of the generator is turned, the rubbing material (woolen cloth) is pressed tight enough to the drum surface to cause friction between the cloth and the drum (Fig. 2-5). This friction causes some electrons from the rubbing material to move to the surface of the plastic drum. The surface of the drum has groups of surplus electrons.

Since the rotating plastic drum is a nonconductor, the surplus electrons leave the plastic and move to the wire brush. The electrons are then conducted to the surface of the metal ball and accumulate there, giving the metal ball a negative electrical charge.

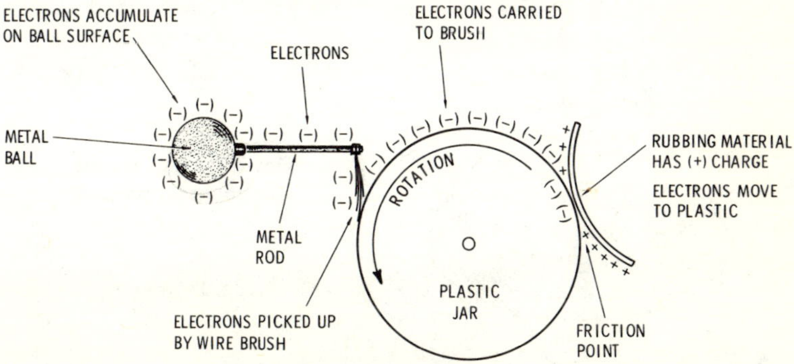

Fig. 2-5. Rotary friction generator theory.

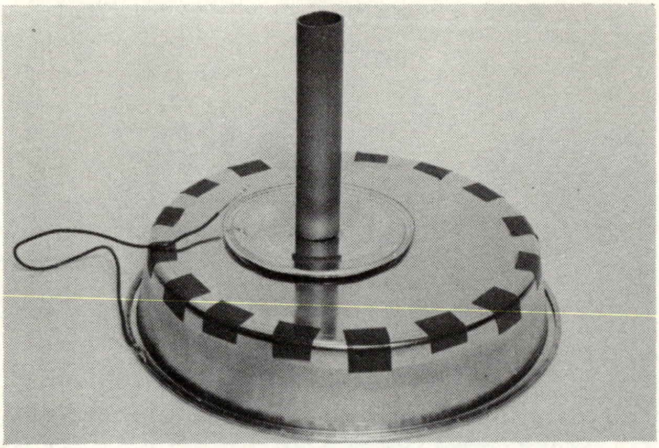

Fig. 2-6. Electrophorus project.

ELECTROPHORUS

Construction of the Electrophorus

Volta's original electrophorus consisted of a slab of resin held in a metal container. It was activated by rubbing it with fur. Our model uses sheet plastic on top of a metal container (Fig. 2-6).

Look at the drawing in Fig. 2-7. Our model of the electrophorus uses a 7½-inch metal pie plate, turned upside down. Any size metal container can be used instead; the size is not critical. Cut a section of thin sheet plastic (sheet acetate was used for our unit) and tape it to the bottom of the metal pan, as shown in the drawing. Use short sections of tape (plastic tape is best) spaced around the pan.

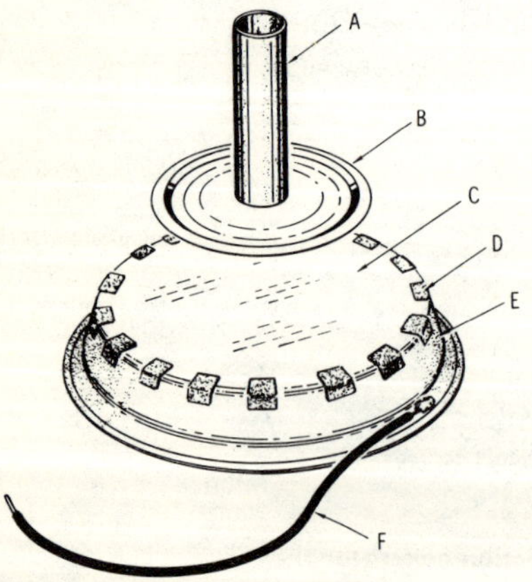

Fig. 2-7. Plans for the electrophorus project.

Solder a length of stranded hookup wire to the side of the metal pan and make sure that the wire is long enough to reach the center of the pan bottom. Remove about 1 inch of the insulation at the free end of the wire. If the metal pan cannot be soldered, use a mounting bolt and nut for the wire.

The charge carrier for the electrophorus is made from 3½-inch metal disc B, cut from a section of tin can, and insulating handle A, made of a plastic rod 1 inch in diameter and 4 inches long. Cement the rod to the center of the metal disc, or solder a flathead screw to the top of the metal disc, and drill and tap a matching hole in the handle.

Materials List for Electrophorus

A—Plastic handle approximately 4" long, 1" diameter fastened to B (See text.) B—Tin-can sheet metal disc, approximately 3½" diameter C—Thin sheet plastic cut to fit top of E	D—Strips of plastic tape to fasten C to E E—Metal pie plate, approximately 7½" diameter (See text.) F—Length of stranded hookup wire attached to E

Electrophorus Theory and Operation

When the plastic sheet on top of the metal pan is rubbed with a cloth (Fig. 2-8A), the plastic surface obtains surplus electrons from the cloth, and the plastic then has a negative electric charge.

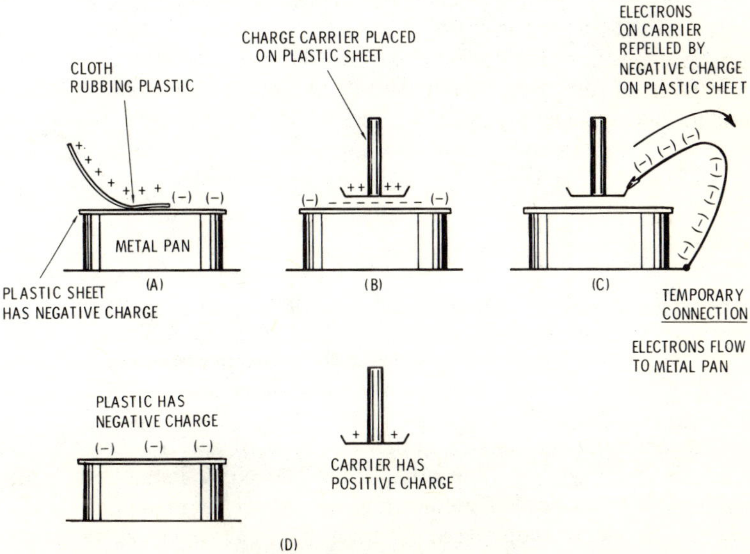

Fig. 2-8. Electrophorus theory.

Fig. 2-8B shows the charge carrier placed on the plastic surface. The negative charge on the plastic then repels the electrons from the surface of the charge-carrier metal disc touching the plastic. When a temporary electrical connection is made between the metal pan and the charge-carrier disc (Fig. 2-8C), the repelled electrons flow to the metal pan. The connection can also be made to ground, instead of to the metal pan.

The electrical connection is then removed and, as shown in Fig. 2-8D, the metal disc of the charge carrier now has a positive electrical charge. The plastic material still retains the negative electrical charge on the surface.

After the charge carrier is discharged, it can be again placed on the plastic surface and electrically connected for a recharge, without rubbing the plastic again. This action can be repeated several times before the plastic will lose all of its negative charge.

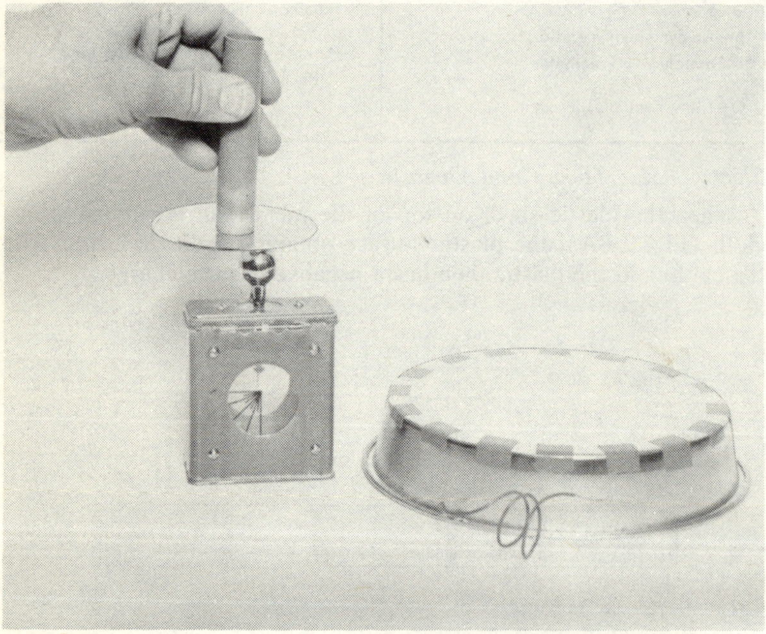

Fig. 2-9. The electroscope shows that the charge carrier has an electrical charge.

After charging the carrier, touch one lead of a NE-2 lamp to the metal disc while holding the other lead. The NE-2 lamp should momentarily light. Also, after charging the carrier, hold the metal disc close to an electroscope (see Chapter 3). The electroscope leaves should move apart, indicating charge (Fig. 2-9).

VARLEY MACHINE

Construction of the Varley Machine

The electrostatic generator invented by Varley is not easy to construct. You will need patience, particularly during the various adjustments of the machine, but the results will be rewarding. In place of the glass discs used by Varley in his original machine, 45-rpm records (approximately 7 inches in diameter) are used in our model (Fig. 2-10). Before starting construction, carefully remove any paper labels that are on the center of these records. If desired, you can use 7-inch discs cut out of sheet plastic that is approximately as thick as a 45-rpm record.

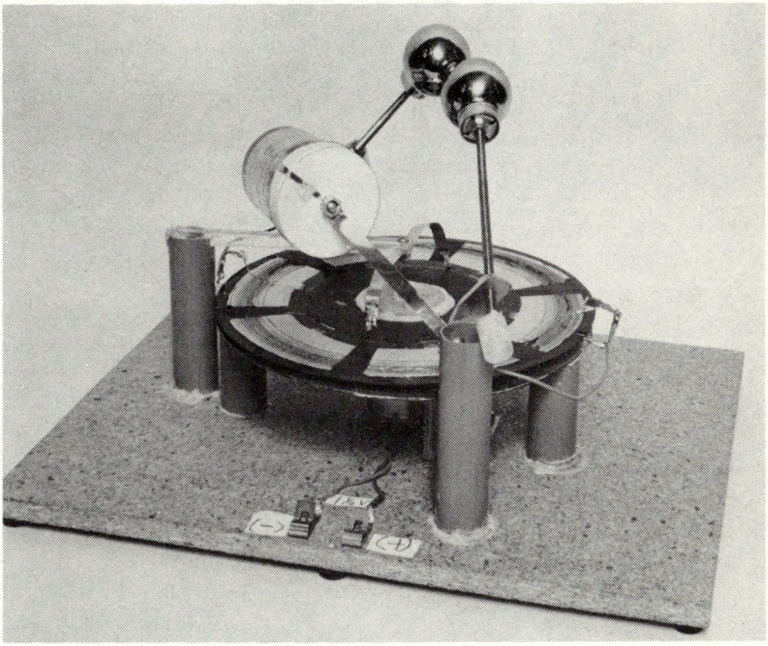

Fig. 2-10. Varley machine project.

Look at the drawing of the top and bottom discs in Fig. 2-11. Cement aluminum foil to one side of each disc; then, using a protractor, lay out and cut the foil patterns with a sharp knife or razor blade, as shown in the drawing. Peel off the foil between the patterns to make the required spaces and temporarily set aside the discs.

The Varley machine is built on a 10 by 12 by ⅜-inch wood or particle board base. For best results, follow the component placement layout drawings (Figs. 2-12 and 2-13) as closely as possible. Electric motor M is a small type with a permanent-magnet field, available in hobby stores or from electronics mail-order houses. The rpm and size of the motor are not critical. Most any size and rpm rating should work, as the only load on the motor is driving top disc J against the friction of brushes H and L. (Our motor has an input voltage range from 1.5 to 4.5 Vdc and a 10,000-rpm rating.)

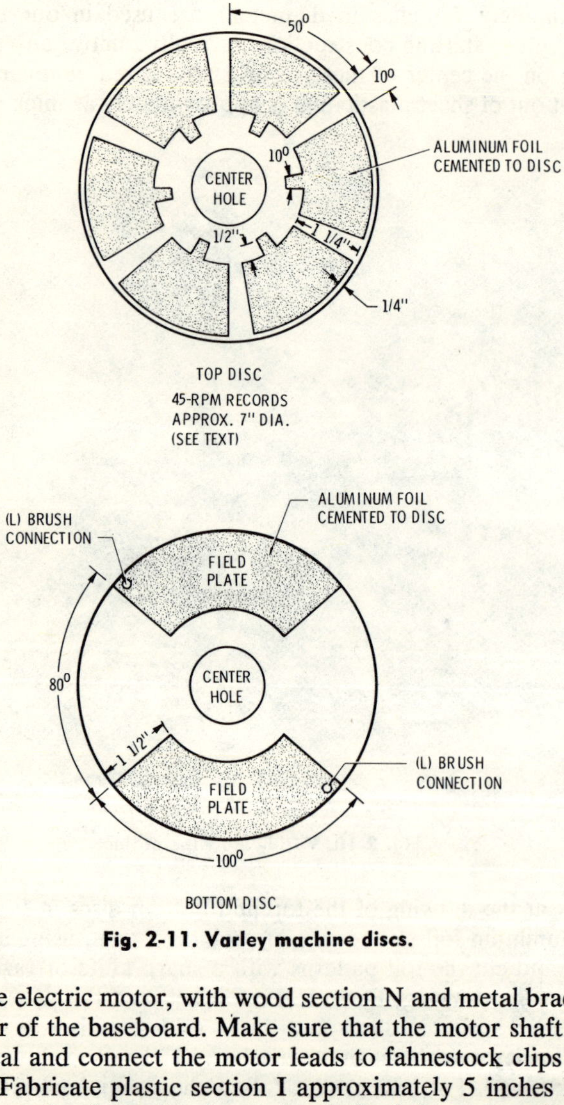

Fig. 2-11. Varley machine discs.

Install the electric motor, with wood section N and metal bracket O, in the center of the baseboard. Make sure that the motor shaft is perfectly vertical and connect the motor leads to fahnestock clips on the baseboard. Fabricate plastic section I approximately 5 inches long to fit the center hole of top disc J and the shaft of the motor. Cut away

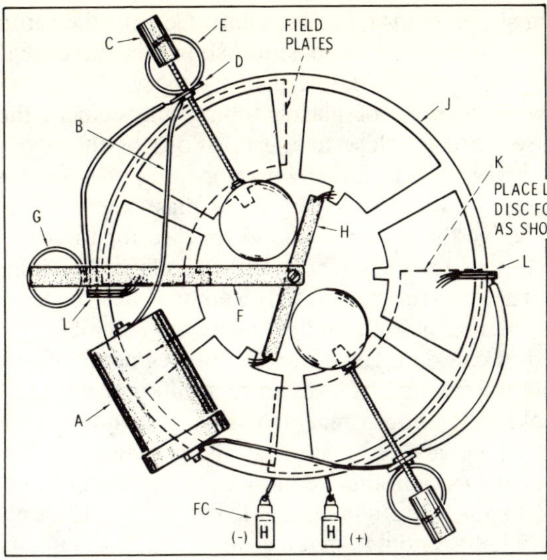

Fig. 2-12. Varley machine assembly, top view.

enough of the diameter of the plastic section to allow it to pass through the center hole of bottom disc K without touching the sides. Drill a center hole in I for the motor shaft, long enough to allow vertical adjustment of the plastic section. Install a setscrew near the bottom of I,

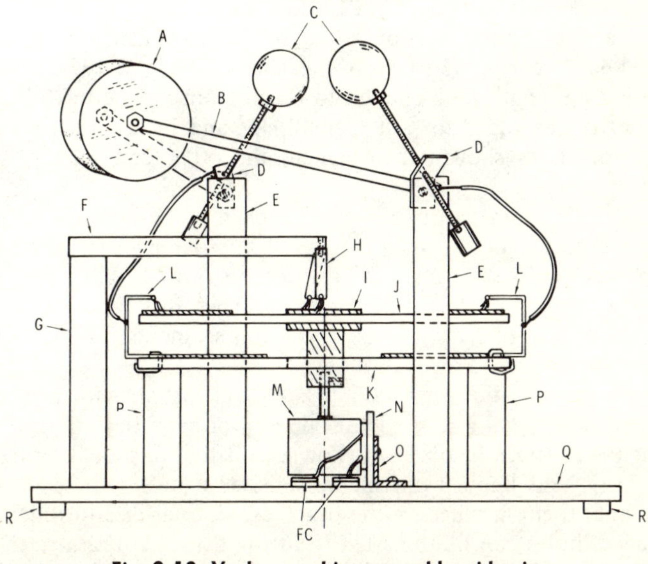

Fig. 2-13. Varley machine assembly, side view.

to keep I from slipping on the motor shaft. Cement the top disc to the top of the plastic section in the position shown in the drawing of Fig. 2-13 and temporarily set aside the disc assembly.

Cut the two P sections of plastic tubing and cement them to the bottom of lower disc K, close to the positions of the two L brushes shown in the drawings. Place the center opening of lower disc K over the motor shaft, and then place top disc J plastic section on the shaft of the motor. Move the center of the lower disc for the best clearance between all sides of the center hole and the plastic section, and then cement the P tubes to the baseboard. Remove the center section and top disc from the motor shaft and set aside the assembly.

The L brushes can be attached to the foil segments of lower disc K at the positions shown in Fig. 2-12 in two different ways. One way is to mount a solder lug with a machine screw and nut in a hole at the foil ends and connect a wire brush, as in the drawing.

To attach the brushes in the second way, begin by drilling two small holes in the foil ends. Carefully bend a length of No. 18 copper wire in the holes and clamp it tight, making sure that the wire has a good electrical contact with the foil. Cut off the wire to make it 1½ inches long and solder a ⅜-inch length of fine-diameter stranded wire to the free end of the solid wire. Repeat for both sides of the foil positions noted in Fig. 2-12.

Install top disc J on the motor shaft. Adjust the space between the top disc and the lower disc to about ¼ inch before tightening the set screw on plastic section I. Bend both L brushes over, so that the stranded wire ends just touch the surface of the top disc.

Cut the plastic tube, section G, to size and cement it in place on the baseboard near the left L brush, as shown in Fig. 2-13. Cut plastic section F to a length long enough to reach from the top of tube G to the center of the top disc. Cut a small slot and install a set screw in the end for H, as shown; then cement the other end on top of the G tube.

Fabricate H from approximately 5 inches of ¼-inch wide tin-can sheet metal and solder ⅜-inch lengths of fine-diameter stranded wires to each end. Insert the center of H into the slot of F; bend down and adjust the ends of H so that the wire brushes just barely touch the surface of the top disc at the tabs of the foil segments. Before tightening the set screw, position H as shown in Fig. 2-12.

Cut the two plastic tubes (E) to size and install, with a machine screw and nut, metal bracket D at one end of each tube (Fig. 2-13). Cement the E tubes in place as shown, and then install the threaded rods, plastic end knobs, and metal balls. Use nuts on the threaded rods to hold them in place on the brackets. For easier adjustment, the top bracket holes can be threaded to fit the threaded brass rods, with a nut used to lock the rod in place.

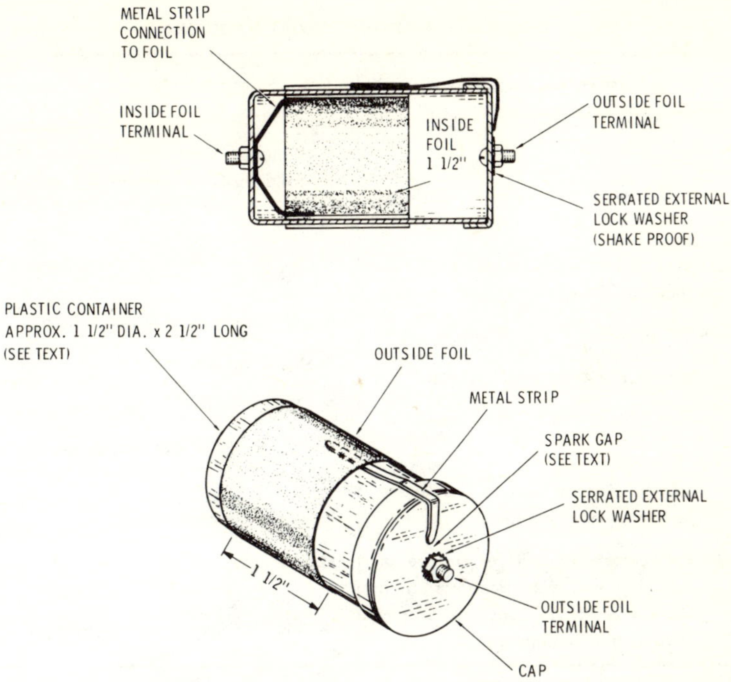

Fig. 2-14. Varley machine capacitor.

Fabricate capacitor A as shown in the drawing of Fig. 2-14. A 2-ounce plastic pill bottle with a plastic screw cap was used in our model. Cut a ⅛-inch wide strip of tin-can metal. Make it pointed at one end and cut the other end long enough to fit under the outside foil strip. (Make sure that there is a good electrical connection.) Use plastic tape to hold the pointed end about 1/32 inch away from the points of the washer of the outside foil terminal screw, as shown in the spark gap of Fig. 2-14. Use two lengths of ¼-inch wide tin-can metal strips (positioned as shown in the drawings of Figs. 2-12 and 2-13) to mount the capacitor to the metal bracket-mounting screws on the E plastic tubes. The capacitor can rest on plastic section F, but make sure that it is away from brushes L and H.

Be sure that the top disc rotates freely and does not touch the bottom disc. If the bottom disc shows signs of sagging, cement additional plastic sections around the bottom to keep the surface of the disc even.

Varley Machine Operation

Make sure that the terminals to the electric motor are marked with the correct polarity so that the battery will not be connected wrong.

Materials List for Varley Machine

A—Capacitor (See capacitor drawing.)	J, K—Top and bottom discs (See disc drawing.)
B—Tin-can sheet metal strips; approximately 1/4" wide	L—High-voltage electrodes bent from solid copper wire with 3/8" stranded wire brushes
C—1 1/4" metal ball with 5" threaded brass rod and plastic end section	M—Electric motor, hobby-type (CER-MAG No. 45 or equivalent)
D—1 1/2" × 1/2" metal bracket with two mounting holes	N—Wood section to fit motor base
E—7/8" diameter × 3 1/4" plastic tube	O—Metal bracket to fit wood section N
F—1/4" square × 5" plastic section	P—7/8" × 2" plastic tube
G—7/8" diameter × 3" plastic tube	Q—10" × 12" × 3/8" particle board for base
H—1/4" tin-can metal strip approximately 5" long with 3/8" stranded wire brushes at both ends	R—Rubber bumper feet
I—Plastic section to fit center hole of top disc and electric motor shaft (cut away for clearance of bottom disc, as described in text)	FC—Fahnestock clips

If the motor rotates in the wrong direction, the wire brushes will bend and dig into the foil, possibly tearing the foil segments.

The electric motor will probably work over a range of input voltages. If so, use the lowest-voltage battery so that the motor will operate at its lowest speed. Connect the battery to the motor terminals and observe that the upper disc rotates freely with all of the brushes just touching the foil segments. Adjust the spark-gap spacing by means of the plastic ends until sparks start to occur. If no sparks are seen, hold an electroscope (see Chapter 3) near the high-voltage brushes; deflection of the electroscope leaves indicates proper operation. Next, adjust the spacing of the pointed spark gap on the capacitor as close as possible (without actual contact). Also try adjusting the position of the central H brushes for best spark operation. (Make sure the brushes touch only the foil segment tabs.) Try touching one lead of an NE-2 neon lamp to one of the spark-gap knobs while holding the other lead. The lamp should light intermittently as the machine operates (Fig. 2-15).

Varley Machine Theory

This electrostatic generator works by induction—a type of rotating electrophorus. Look at the schematic drawing of the Varley machine (Fig. 2-16). The two field plates have the same function as the charged plastic sheet of the electrophorus project, but one field plate will have a negative charge and the other plate will have a positive charge.

Assume that there is no initial electrical charge on the field plates. When power is applied to the electric motor, causing the rotor to move in clockwise direction, the motion of the air past the field plates will cause a very small electrical charge to occur on each plate. The electrical charge will be different on each plate; one plate will be negative and the other plate will be positive.

As each of the six rotating foil segments move past a field plate, an electrical charge will be formed in the segment by induction. The center set of fixed H brushes connect two foil segments, so that surplus electrons will flow between them. Therefore, one of the foil segments will be charged positively and the other will be charged negatively. Each of these segments will then be charged (by induction) to a polarity opposite that of its field plate.

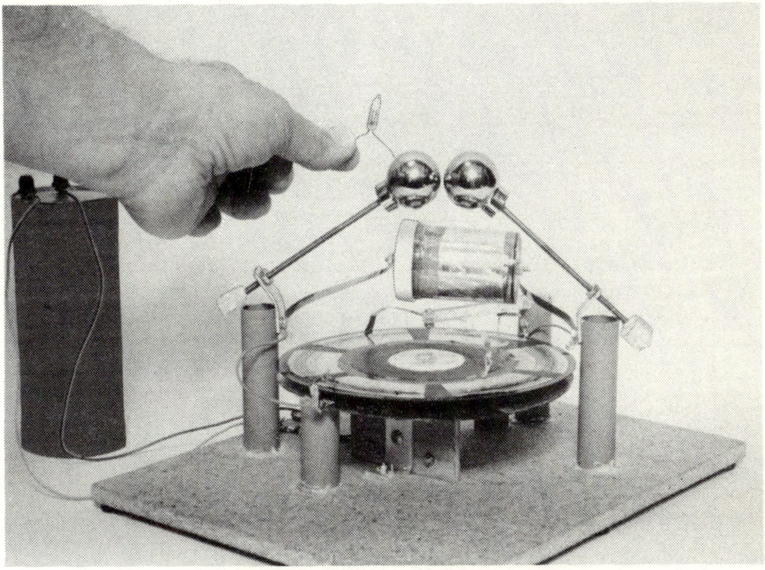

Fig. 2-15. Testing the Varley machine with a neon lamp.

The rotation of the disc will then bring each charged field plate to a high-voltage brush (L), and the electrical charge will then be transmitted through the brush to its field plate. This process is repeated with each of the six foil segments each rotation, increasing the electrical charge on the field plates. As the electrical charges increase on the field plates, the induced electrical charges of the rotating foil segments will also increase. Therefore, in a short time, the electrical charges will rise to a very high voltage.

The capacitor is connected, through a small internal spark gap, between the field plates. This spark gap charges the capacitor by a small electrical leakage (corona discharge) across its points.

When the capacitor is charged up to a high voltage, it will discharge through its internal spark gap and across the large spark gap (formed by the two metal balls).

The rotation of the disc then repeats the whole process of induction; charging the field plates and capacitor until electrical energy is accu-

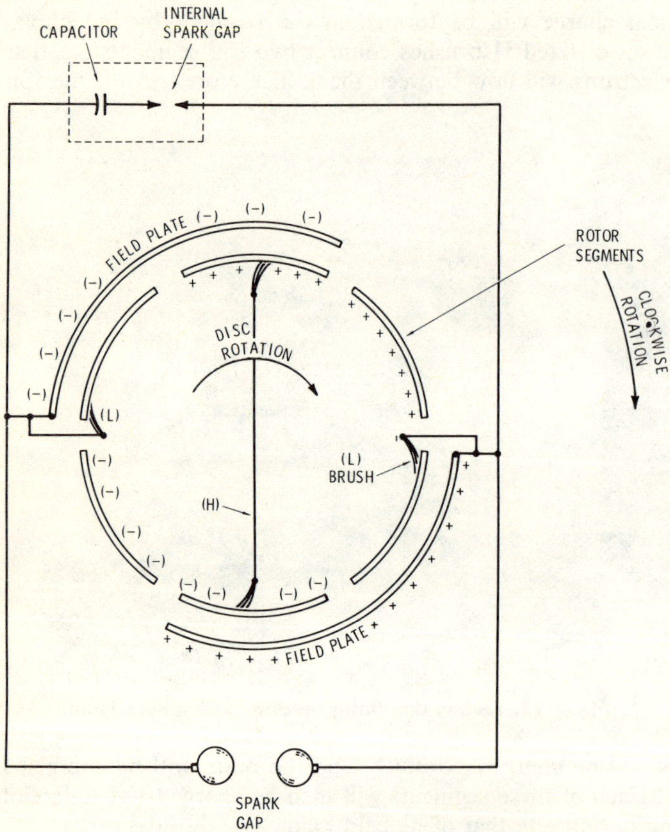

Fig. 2-16. Schematic of the Varley machine.

mulated at a voltage high enough to discharge through the spark gaps. The spacing of the metal balls determines the rate at which the charging and discharging take place. The farther apart they are, the longer it takes between sparks. If they are too far apart, no sparks will occur; electrical losses will discharge the capacitor and field plates at a voltage determined by resistance, corona losses, and system efficiency.

THE WIMSHURST ELECTROSTATIC GENERATOR

The electrostatic generator (Fig. 2-17) invented by James Wimshurst is the most efficient of the rotating-disc inductive generators of the 19th century. This generator is still used today, primarily in school lecture demonstrations. In scientific use it has been superseded by the Van de Graaff generator.

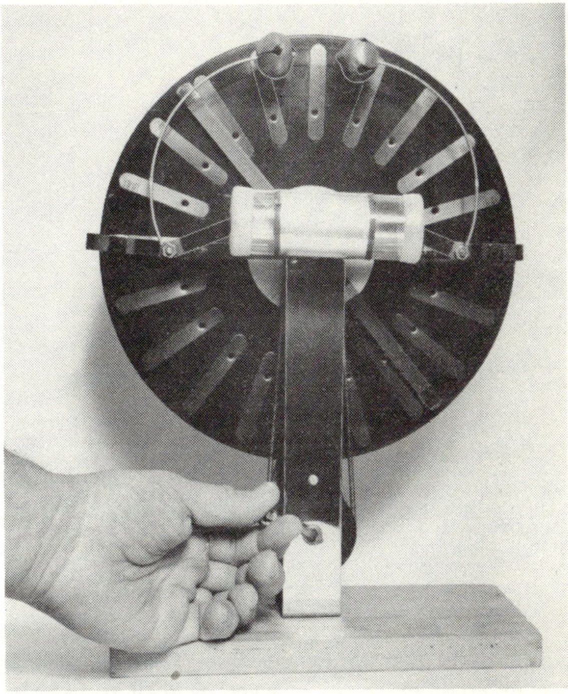

Fig. 2-17. A typical Wimshurst electrostatic generator.

This generator is considered to be too difficult for the average experimenter to construct, and the Varley machine theory and operation is similar enough for demonstration. Wimshurst machines can be obtained from science supply stores and mail-order houses.

As shown in the schematic drawing of Fig. 2-18, this generator uses two discs rotating in opposite directions. These discs are usually made of glass. Modern versions may use plastic discs or phonograph records. The Wimshurst machine is built in a variety of forms and sizes, some with multiple rotating discs, but the basic theory of operation is the same. Most of the machines are rotated by hand cranks.

The rotating discs have a number of metal foil segments, placed near the disc edges, which act both as electrical charge carriers and

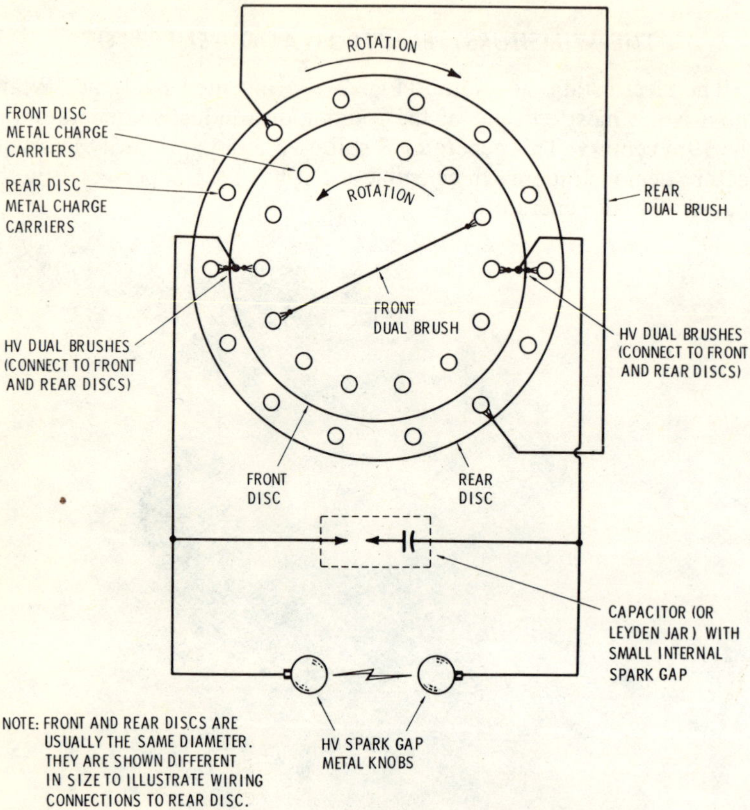

Fig. 2-18. Schematic of typical Wimshurst generator.

as the field plates of the Varley machine. The rotating foils on each disc induce different electrical charges as they pass by each other. The center, fixed, dual brushes (on each side) connect opposing foils. The HV dual brushes connect the high voltage charges to the capacitor and spark gap, as explained in the Varley machine theory of operation.

KELVIN WATER-DROP GENERATOR

Construction of the Kelvin Water-Drop Generator

The Kelvin water-drop generator is pictured in Fig. 2-19. Begin construction by fabricating the water-drop generator stand (Fig. 2-20). Cut the two hardboard (or plywood) 8½- by 12-inch B sections from 1/16-inch thick (or heavier) stock. Cement them to the four, 1 inch diameter by 24 inches long E plastic tubes in the positions shown in Fig. 2-21 (one tube at each corner).

The dimensions of metal pan A are not critical. A baking pan 5½ inches wide by 9¼ inches long by 2½ inches high, made of plated sheet iron was used in our model. Two metal tubes are soldered in holes at the front corners as shown.

The two plastic eyedroppers, G, can be obtained at most drug stores. Try to find the type with as small an opening as possible. Cement the eyedroppers to the ends of the 3½ inches long by ¼-inch square plastic F sections which are cemented about 1 inch down from

Fig. 2-19. Kelvin water-drop generator project.

the top of the two front E plastic tubes as shown in the drawing of Fig. 2-20.

If necessary, cement plastic tubes in the tops of the eyedroppers to fit the short D sections of rubber tubing between the metal tubes, C, of pan A. The exact length of the rubber tubes is not critical; just make them long enough to allow placement of the metal pan in the center of top section B.

The 1½ inch diameter by ⅜-inch wide metal H rings can be made of any metal that can be soldered. We used sections of brass tubes in our model, but tin-can metal rings will work. Solder a length of solid copper wire to the sides of the metal rings and attach them to the front of the E plastic tubes (about 4 inches down from the end of the eyedroppers) by winding several turns of the solid wire about the plastic tubes. Adjust the ring positions so that the centers of the rings are directly below the centers of the openings in the eyedroppers.

Position the two metal cans, J, about 3 inches apart on top of the two plastic or glass M jars; then connect the wires between the rings and the metal cans, as shown in the drawing. Keep the wires spaced well apart and as high as possible from the base. The wires can be

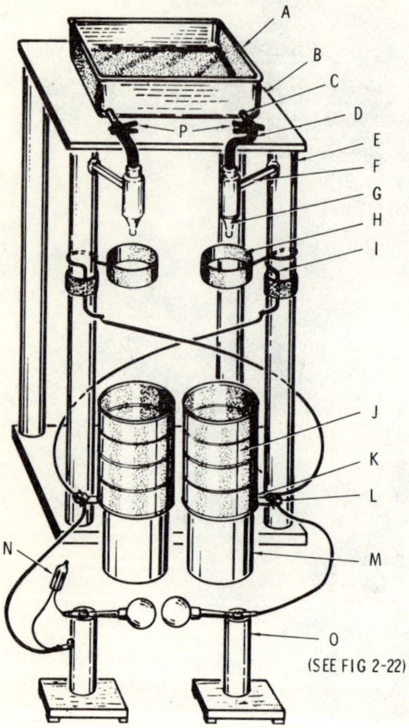

Fig. 2-20. Plans for Kelvin water-drop generator.

connected with alligator clips to small tabs soldered on the metal cans, or they can be connected to metal discs fastened to the top of each jar. This will allow the cans to be picked up and emptied without disconnecting any wires. The electrical connections will then be made between the metal discs and the can bottom.

Fabricate the two spark-gap assemblies as shown in (Fig. 2-22). The two are similar in construction, except that one assembly has neon lamp N mounted on it. Connect the two spark-gap assemblies to the metal cans and space the metal balls about ¼ inch apart. Keep the wires to the spark gap as high from the base as possible. Install two

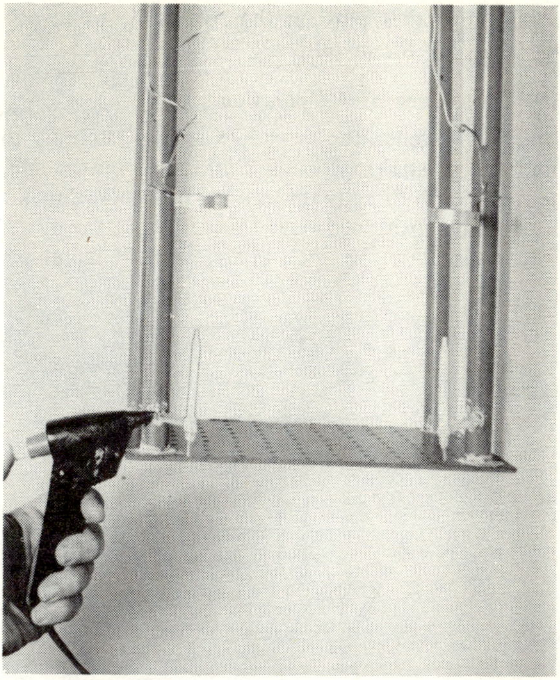

Fig. 2-21. A hot glue gun is used to cement the Kelvin water-drop generator.

Materials List for Kelvin Water-Drop Generator

A—Metal pan approximately 5½″ wide × 9¼″ long × 2½″ high (See text.)	I —Solid copper wire (soldered to H)
B—1/16″ hardboard (top and bottom sections) 8½″ × 12″	J —Metal can, 4″ diameter × 5½″ high (1-lb coffee can or equivalent)
C—1/16″ diameter metal tubing (soldered to A)	K—Metal tabs soldered to J
D—Rubber tubing to fit over C	L—Hookup wire with alligator clips to attach to K
E—Plastic tubes approximately 1″ diameter × 24″ long	M—Plastic or glass jar approximately 2½″ high with a diameter to fit under J
F—¼″ square plastic section 3½″ long	N—¼ watt neon lamp, NE-2 or equivalent (See Fig. 2-22.)
G—Plastic eyedropper	O—Spark-gap assembly (See Fig. 2-22.)
H—Metal ring, 1½″ diameter × ⅜″ (See text.)	P—Plastic or wooden clothespins

wooden or plastic clothes pins on the rubber hoses to act as water flow shutoff valves for the metal pan.

Kelvin Water-Drop Generator Operation

Set up the Kelvin generator as shown in the assembly drawing of Fig. 2-20 and clip plastic or wooden clothespins on the rubber tubes. Position the metal cans directly under the H rings and make sure that the eyedroppers are positioned over the center of the rings. Position the spark-gap balls about ¼ inch apart, and fill metal pan A with water.

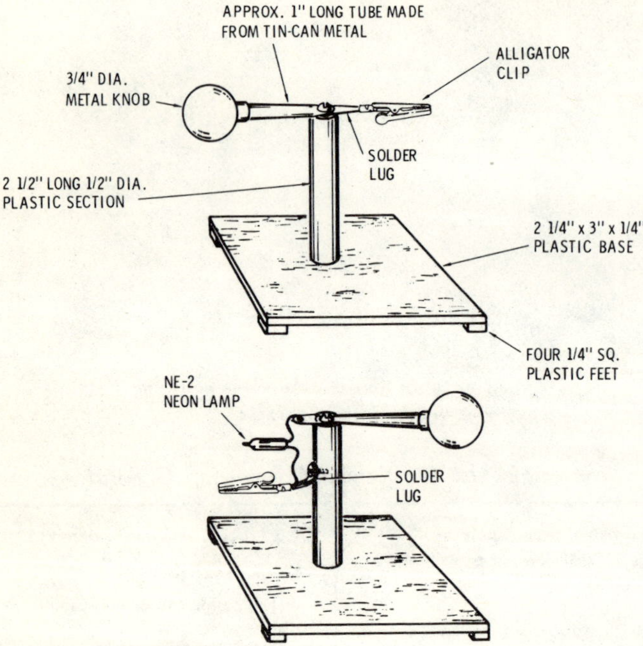

Fig. 2-22. Spark-gap assemblies used with the Kelvin water-drop generator.

Remove both clothespins from the rubber tubes at the same time; if necessary, adjust the wire supports for the metal rings so that the streams of water fall directly in their centers. Move the spark-gap metal balls closer together until sparks are observed. There will be periods of time between the sparks because of the time it takes to build up charges in the metal cans.

After the water has been drained from the top container, check the charges on the cans and container with an electroscope (see Chapter 3). You should find different charge polarities for each metal can. Put the clothespins in place on the rubber tubes, discharge the metal cans by touching the spark-gap knobs together, and fill up the top container with water from the metal cans to repeat the operational cycle.

Kelvin Water-Drop Generator Theory

This type of generator uses the induction principle. In our model of the water-drop generator, the water is in a metal tank insulated from the ground, and the two metal containers, X and Y, are also insulated from each other and ground. As shown in the drawing of Fig. 2-23, there are two streams of water drops. As the water drops fall through the air, they obtain a small electrical charge from friction with the air molecules.

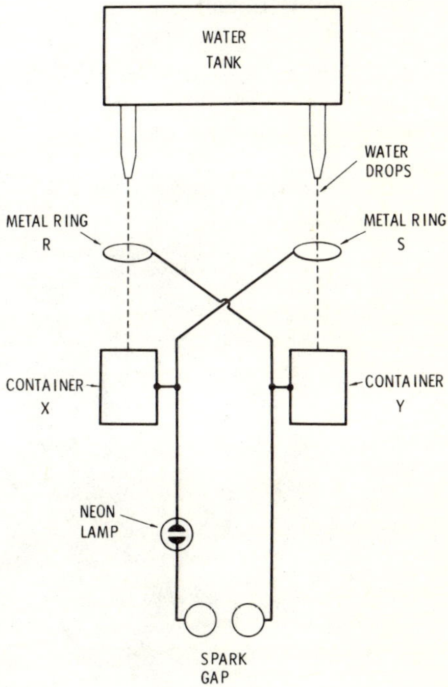

Fig. 2-23. Schematic of the Kelvin water-drop generator.

The small electrical charges of each stream of water are different; one stream will become more negative than the other. As the water drops hit the container, their electrical charge is conducted to the metal container. Thus, each container will have a small electrical charge; one container will be negative, and the other container will be positive.

As the water drops continue to fall through metal rings R and S, an opposite polarity charge is induced in each metal ring. Then, as each ring is connected to the container of the opposite stream, the charge of that container is accordingly increased. Thus, the total charge potential between the metal containers is increased to a high voltage.

When the voltage increases to a high enough point, a spark jumps between the metal balls of the spark-gap assembly and the neon lamp lights, discharging the system.

The charge process repeats and the spark gap and neon lamp operates; charging time depends upon the spark-gap distance. This will continue until the water flow stops.

VAN DE GRAAFF GENERATOR

Construction of the Van de Graaff Generator

The Van de Graaff generator usually has a large metal globe. A metal globe may be difficult to find; so our model is designed to use an aluminum beer can with a rounded end on the bottom (Fig. 2-24).

Fig. 2-24. Van de Graaff generator project.

(We used 12-ounce beer cans for our model.) With sandpaper, remove all the paint from the outside of the can and make sure that the inside of the can is clean and dry. The exact size of the can is not critical, but it should be a minimum of 2½ inches in diameter.

The rubber belt used in our model is a large rubber band approximately ⅜-inch wide and 6 inches long. The size is not critical; a rubber belt can be cut from an inner tube or from a section of rubber sheet, cemented together. Make sure that the rubber belt is clean and dry.

Handle it very carefully to avoid getting it oily or moist from your hands.

Look at the drawings of Figs. 2-25 and 2-26. The generator is built on a 6- by 10- by ⅜-inch section of particle board (wood or plywood can also be used). The electric motor is the same small hobby type used for the Varley machine. The motor pulley is a section of plastic rod approximately ⅜ inch in diameter and ⅝ inch long.

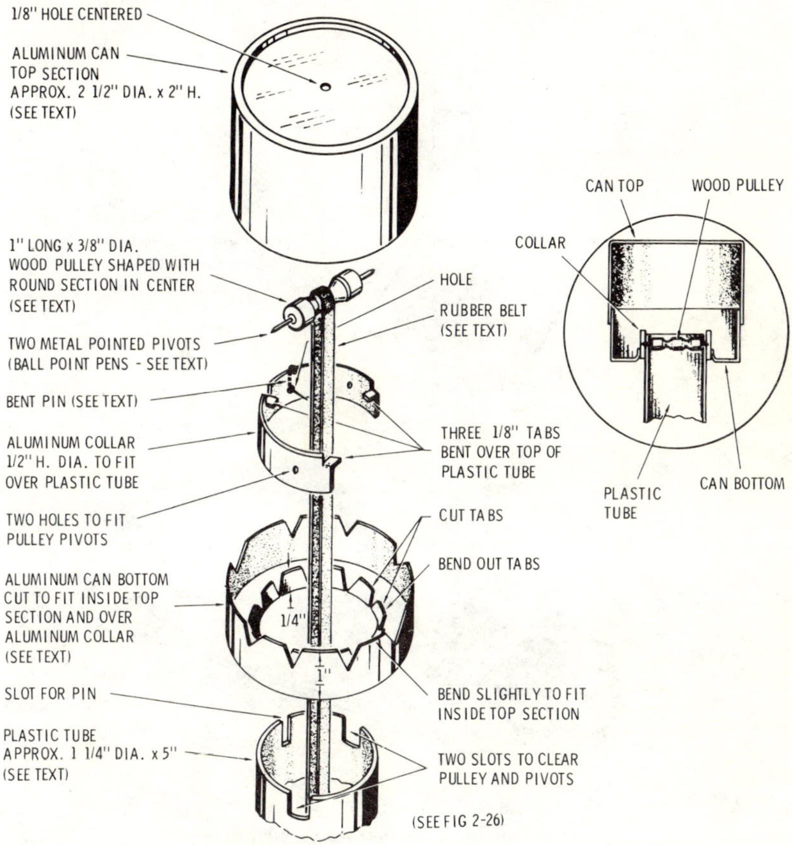

Fig. 2-25. Van de Graaff generator top assembly.

Groove the center of the pulley to keep the belt in place. Drill a center hole in the end of the plastic to fit the motor shaft and mount it on the shaft with a short set screw to hold it in place. Make sure the set screw is short enough so the outer end of the set screw is below the surface of the plastic.

Mount the motor on the board in a position such that the plastic pulley is at the center of the board. Cut a 3-inch-square plywood platform and cut a center hole to fit the plastic tube as shown in the drawing. Cement the plywood platform with its hole center directly over the motor pulley with wood dowels at each corner to clear the motor housing. Connect the motor leads to terminal clips mounted on the board.

The length of the plastic tube is determined by the length of the rubber belt. In our model, the plastic tube is approximately 5 inches long by 1¼ inches in diameter. Select a tube length a little longer than

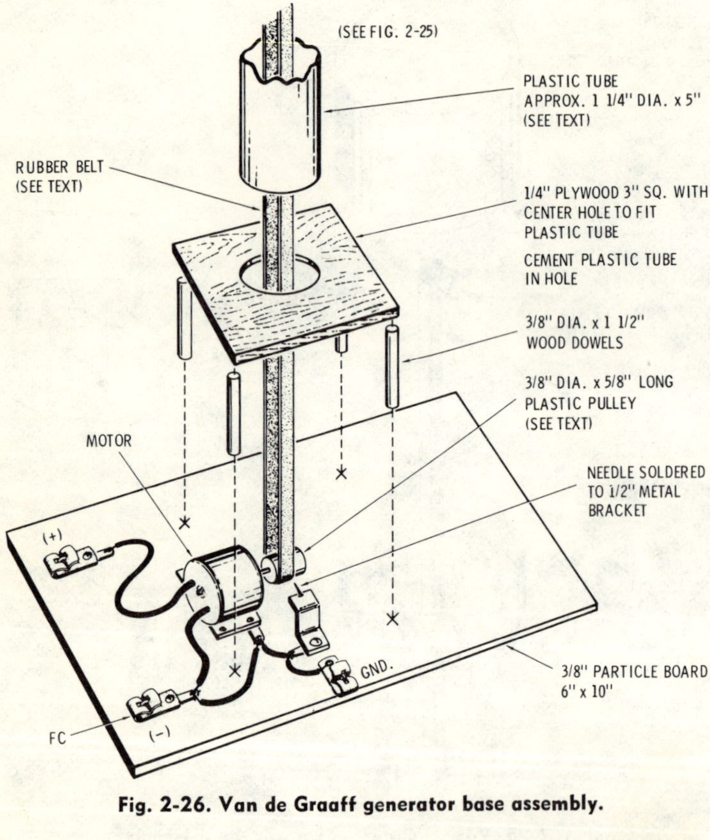

Fig. 2-26. Van de Graaff generator base assembly.

you think will be required; the tube can be cut to size after final assembly. Cement two short sections of wood under the plywood platform, extending about the thickness of the tube walls into the edge of the center hole. This will allow the plastic tube to rest in the hole while adjustments are made.

Cut the aluminum top section to size from the rounded beer can bottom, as shown in the drawing of Fig. 2-25, and set it aside. Fabricate the aluminum collar of heavy-gauge sheet aluminum to fit around the outside of the plastic tube. Cut the three tabs and bend them to keep the collar in place. Set the collar aside.

Make the wood pulley, as shown in the drawing, with the center section rounded to keep the rubber belt from slipping to one side. We used two ball-point pen tips from the pulley pivots (a metal shaft pointed at each end will also work). If you use ball points, make sure that they are from pens that no longer contain ink. Wash the pen points carefully in alcohol or solvent to remove all of the ink and dry them carefully. Make sure that the pens are inserted in the exact center of the wooden pulley so that it will rotate freely without wobbling. Drill two small holes in the aluminum collar, located as shown in the drawing, and push the pen points into the wood pulley, allowing enough clearance for the pulley to rotate freely in the collar.

Place the plastic tube in the hole of the platform. Put the rubber belt in the tube and over both the wood pulley and the plastic motor pulley; make sure it is not twisted. Locate and cut the clearance holes at the top of the plastic tube for the wood pulley and pin. Locate and drill the hole for the pin so that the pin point will be close to the belt. Then, bend the pin over the top of the collar to hold it in place. Make sure that the pin point is close to, but not touching, the rubber belt.

Cut the bottom section of aluminum, as shown in the drawing of Fig. 2-25, from another rounded beer can bottom. Cut out the bottom section center hole and bend the tabs so that the bottom section will fit snugly over the aluminum collar. Cut out the top tabs and bend them in a little so that the top section will fit snugly over the bottom section.

Check the rubber belt position and tightness. It should not be taut but should fit over the two pulleys in a straight line. Rotate the pulleys; they should smoothly move the rubber belt with little effort. If necessary, the length of the plastic tube should be changed for the best fit of the rubber belt before cementing the tube in the platform hole.

Materials List for Van de Graaff Generator

Wood stock for base	Plastic tube, 1¼″ diameter × 5″ long (See text.)
¼″ plywood and ⅜″ diameter wood dowels for platform	Aluminum can, approximately 2½″ diameter × 3″ (minimum) high (See text.)
Electric motor (See text.)	⅜″ wood dowel for pulley (See text.)
Sheet aluminum for brackets	⅝″ plastic rod for pulley (See text.)
Fahnestock clips for terminals	Pins for electrodes
Rubber belt (See text.)	
Cement and screws to mount components	

Solder a needle or pin to a ½-inch metal bracket and position it as shown in the drawing of Fig. 2-26. The pin point should be close to, but not touching, the rubber belt on the plastic pulley. Fasten the bracket on the board with a solder lug connected to a ground clip. A lead should also run to the motor housing and one to the motor terminal clips.

Van de Graff Generator Operation and Experiments

Connect a battery to the electric motor terminals and make sure that the motor drives the belt upward on the side between the ground needle electrode and the can needle electrode. If not, reverse the battery connections and mark the motor terminals. Check to see that the two needle points are not touching the rubber belt. Position them so that they are as close as possible to the center of one side, but are not touching the belt. Connect a lead from an external ground to the GND terminal on the baseboard.

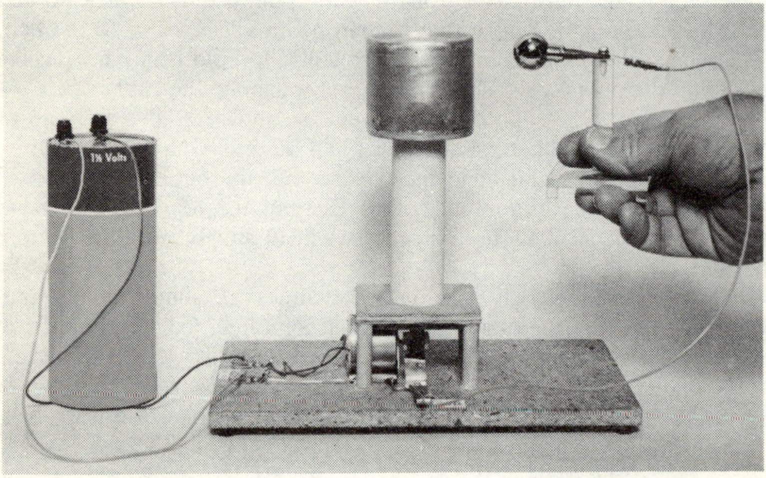

Fig. 2-27. Testing the Van de Graaff generator with a spark-gap assembly.

Connect a 24-inch lead to the GND terminal (do not disconnect the external ground lead) and connect the other end of the lead to a metal ball (Fig. 2-27). Hold the metal ball near the side of the can and observe the sparks. Gradually move the metal ball outward, away from the can, until you find the maximum distance that the sparks will jump (our model had 1-inch sparks). You can also use the spark-gap assembly unit with the neon lamp designed for the Kelvin generator. Disconnect the metal ball, and connect the neon lamp ball unit in its place. Put the neon assembly on a plastic support (such as the plastic experiment table of Chapter 7) that will bring the metal ball of the neon lamp assembly to approximately the center of one side of the Van de Graaff generator can. Adjust the spacing between the metal ball and the can for the desired spark rate and observe that the neon lamp flashes each time that the spark operates. The time between sparks is the time required to build up the charge in the can. This time is dependent upon the spacing of the spark gap. The greater the spark-gap distance, the longer the generator will take to build up the required high voltage. (Small spark-gap spacing involves lower voltages and, therefore, takes less time to build up.)

Move the spark-gap assembly (or an NE-2 neon lamp) around the can outside the spark operating range. (This should be done in a darkened room.) Observe that the neon lamp glows a distance away from

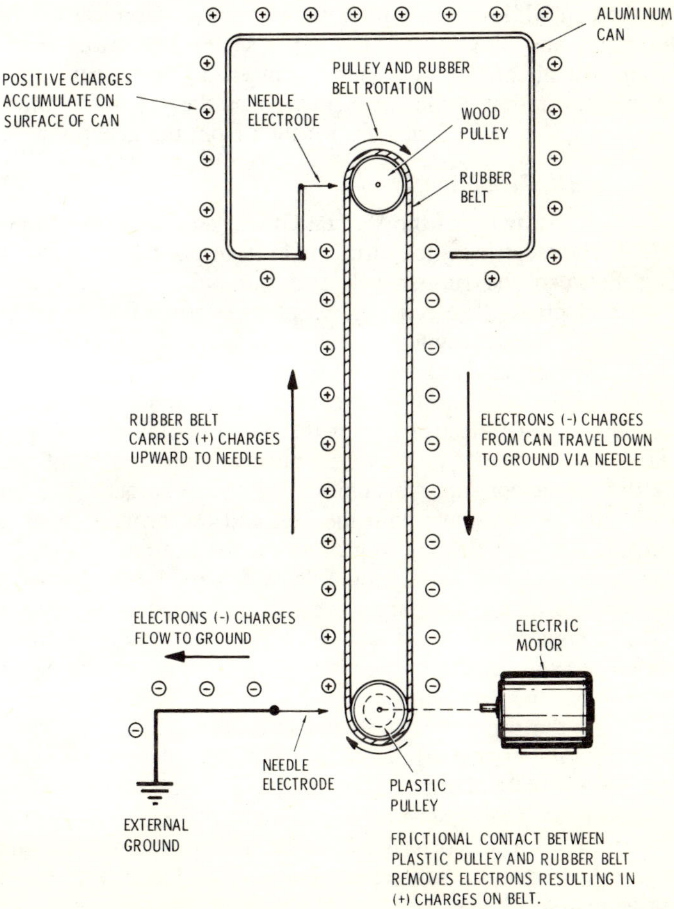

Fig. 2-28. Schematic of Van de Graaff generator.

the can, mapping the extent of the electrostatic field surrounding the can.

Observe the generator operation with the room totally dark and you may see faint streamers of light emitted from the can. This is due to corona discharge—electrical charges striking moisture and dust particles in the air surrounding the can. These corona discharges are charge losses for the generator, and they can be minimized on the bottom half of the can by taping the bottom edge with a layer of plastic electrical tape. The corona discharges have more of a tendency to jump from this edge toward ground.

Cement or tape several sections of 1½ inches long by ⅛-inch wide tissue paper to the head of a pin or ¾-inch length of No. 22 solid wire. Place the end of the wire or pin into the hole in the center of the top of the can while the generator is *inoperative.* Connect the battery to the motor and observe that the paper sections fly away from each other and lift off of the can. This is caused by the paper gaining an electrical charge that is the same polarity as the generator. Since like charges repel, the paper strips are repelled from the generator.

Van de Graaff Generator Theory

Look at the drawing of the Van de Graaff generator operation (Fig. 2-28). As the electric motor turns the bottom plastic pulley, frictional contact between the rubber belt and the pulley generates a small amount of electrostatic electricity. Surplus electrons flow to the needle point and to ground, leaving the belt with areas of electron deficiency (positive electrical charge).

These positive electrical charge areas are carried up as the belt moves, and surplus electrons from the metal container on top of the belt flow, via the needle point, to the positive charge areas of the belt. Thus, the metal container becomes positively charged. The surplus electrons are moved down with the belt and the process repeats with the surplus electrons flowing to ground via the bottom needle.

The top metal container accumulates a very large positive high-voltage charge and it discharges when a grounded object is brought near. Also, small brush electrical discharges (corona discharges) are taking place, with electrons flowing to the positively charged metal container from the nearby air molecules.

Each time the top metal container is discharged, the electron flow between the rubber belt and container is repeated until the container becomes electrically charged again. If the plastic pulley were on top and the wooden pulley were on the bottom, the electrons would flow from the ground to the belt and from the belt to the metal container; thus, giving the metal container a negative electrical charge instead of a positive charge.

CHAPTER 3

Electrostatic Detectors

The early experimenters did not have the multitude of electrical meters and devices available to modern-day experimenters and scientists. (Fig. 3-1). The early experimenters tested the substances, electrified after rubbing, by their ability to attract lightweight items.

PITH-BALL INDICATING DEVICES

The first electrical indicating instrument, the "versorium," invented by Gilbert, was a great step forward in testing electrified substances.

Fig. 3-1. Some of the electrostatic detector projects included in this chapter: (left to right) gold-leaf electroscope, shielded electroscope, electrostatic voltmeter, twin-electrode foil electroscope.

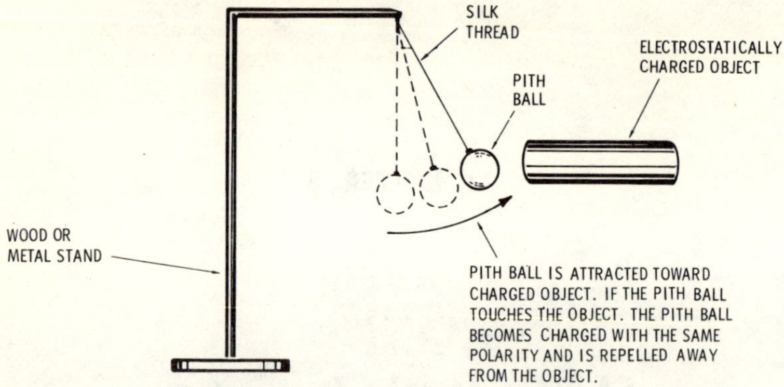

Fig. 3-2. Single pith-ball indicating device.

A single pith ball suspended on a silk thread was also used by early experimenters, as shown in Fig. 3-2. Pith balls were made from the dried inside stem of the elderberry. Cork and lightweight metal sections were also used.

Another advance in instrumentation was the utilization of the repulsion of two similarly charged items. Benjamin Franklin used an electrostatic indicator employing two linen threads hanging together. Other experimenters used dual straw sections or pith balls hanging close together (Fig. 3-3). When both the items (linen threads, straw sections, pith balls, etc.) had similar electric charges, they would repel each other. The distance or angle of repulsion would give a rough idea of the strength of the electrostatic charge. Later experimenters housed dual charged items in glass jars, with electrical connections made to a metal ball on top of the jar. This assembly is called the "electroscope." In 1787, Abraham Bennet substituted gold leaves for pith balls or threads, and the electroscope was further evolved into a sensitive instrument used by scientists in modern times.

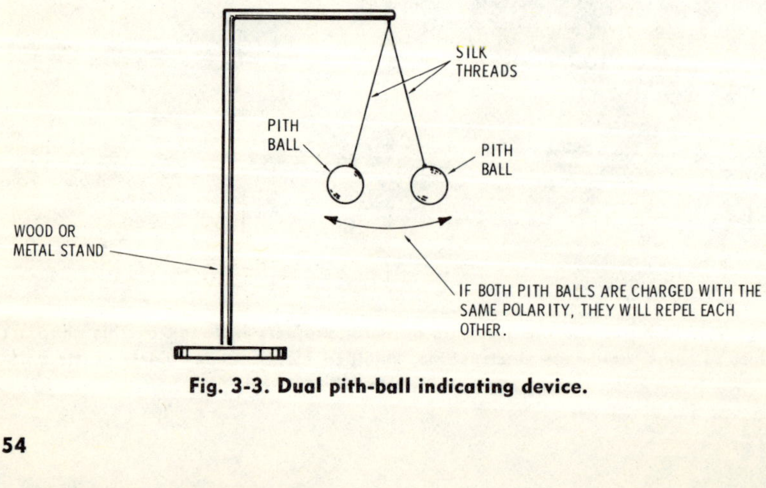

Fig. 3-3. Dual pith-ball indicating device.

This chapter covers indicating-device projects, from an electroscope to an electrostatic voltmeter, that can be used for experiments throughout this book.

VERSORIUM

Construction of the Versorium

Gilbert's oirginal versorium had a pivoted, light, metal bar, but our model uses a drinking straw with aluminum foil cemented on it (Fig. 3-4).

Fig. 3-4. Versorium project.

Begin construction by inserting the head of a pin or needle into the center of a 2-inch-square block of balsa wood or any other type of soft wood. The balsa wood can be obtained at any hobby shop that sells model airplane kits. Make sure that the sharp end of the pin or needle is vertical (Fig. 3-5).

Cut a 2-inch section of a paper or plastic drinking straw, and cut an opening in the middle of the straw that will allow the pin to pass through into the center of the straw. Coat the straw with cement and roll it on a sheet of household aluminum foil. Bend the foil around the straw and cut off the excess foil.

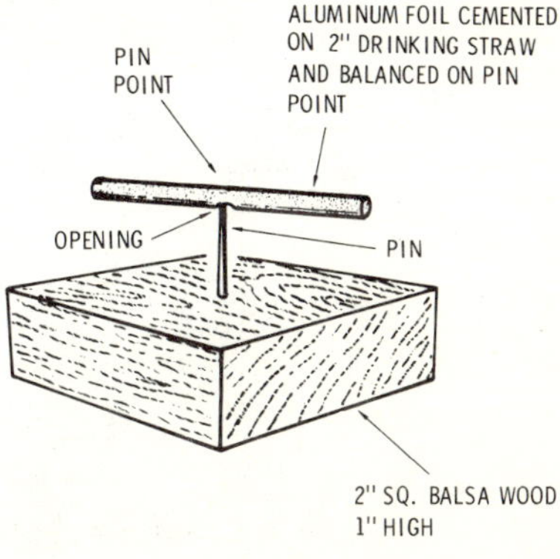

Fig. 3-5. Plans for versorium project.

Insert the pin or needle through the opening in the middle of the straw and balance the straw on the needle or pin point by *lightly* pressing the point against the inside surface of the straw.

Gently rotate the straw and make sure that it is balanced on the point of the pin or needle. If necessary, cut off an end section to balance the straw.

Materials List for Versorium

Aluminum foil	Pin or needle
Drinking straw	Balsa wood

Versorium Theory and Operation

Our version of the versorium is constructed of a straw covered with metal foil, and it operates by induction, as shown in Fig. 3-6. When a charged object is placed close enough to the versorium so that its metal foil is within the electrostatic field of the object, the versorium rotates toward the charged object.

If the object has a positive electrostatic charge, the surplus electrons of the foil try to move closer to the positive charge. This action swings the versorium toward the charged object.

If the object has a negative charge, foil surplus electrons try to move away from the object (to the extreme end of the foil, away from the

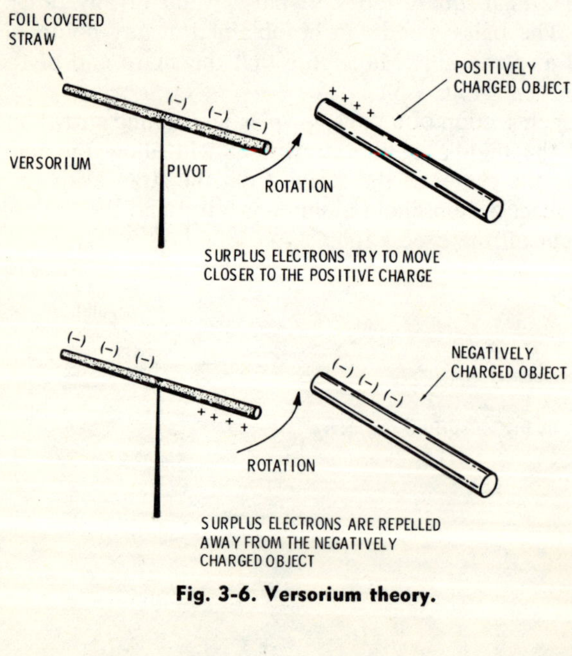

Fig. 3-6. Versorium theory.

object). The resultant positive charges left on the foil end closest to the object rotate the versorium toward the negatively charged object.

Make sure that the versorium is balanced properly and rotates freely. Then, rub a plastic rod with a woolen cloth and hold the rubbed (electrically charged) end near the versorium. The versorium should be attracted toward the charged rod. Try the same experiment with a rubbed glass rod; it should work the same way, showing that the versorium will operate with either polarity of charge.

GOLD-LEAF ELECTROSCOPE

Construction of the Gold-Leaf Electroscope

The original electroscope used by early experimenters had leaves made from very thin sheets of hammered gold. These thin sheets of gold are called gold foil or gold leaf. In our model, we used simulated gold leaf (simulated copper leaf can also be used), obtainable at art

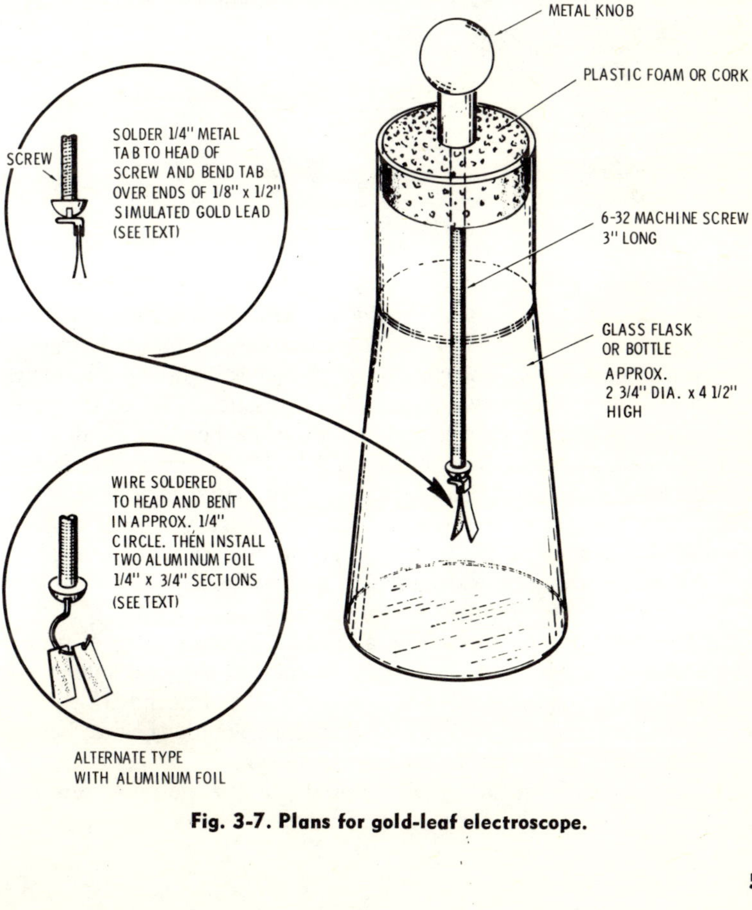

Fig. 3-7. Plans for gold-leaf electroscope.

supply stores. This type of material is inexpensive, electrically conductive, and, being very thin, makes a very sensitive electroscope. The simulated gold leaf may be difficult to work with because it is extremely thin and fragile. The drawing of Fig. 3-7 shows an alternate type of construction using household aluminum foil for the electroscope leaves. Aluminum foil is easier to work with, but it will not be as sensitive as the simulated gold leaf.

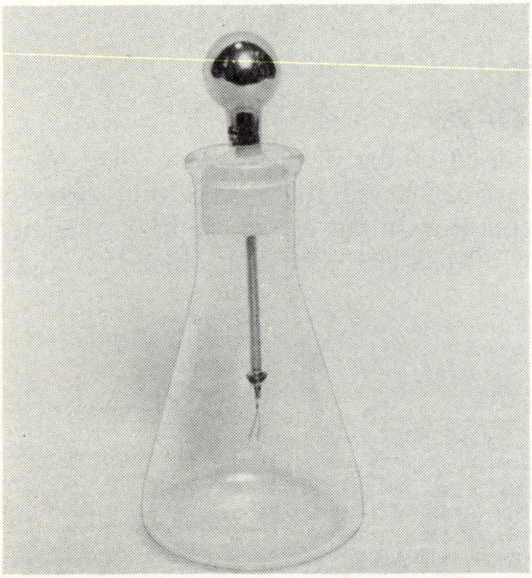

Fig. 3-8. Gold-leaf electroscope project.

The electroscope is built in a glass bottle (Fig. 3-8). A bottle of any size or shape can be used, provided it will allow enough space for free movement of the foil leaves. Our electroscope model is built in a laboratory glass flask approximately 4½ inches high, and 2¾ inches in base diameter. As shown in Fig. 3-7, a plastic (solid or foam) or cork stopper is used in the electroscope. Plastic is best electrically (less electrical leakage than cork) and should be used if available. If cork or some other material is used for the stopper, slip a section of plastic tubing over the machine screw for better electrical insulation. If a different size of glass bottle is used than the one we used, change the length of the machine screw to a length that will extend approximately halfway into the glass bottle.

Start construction by cutting out a ½- by ¼-inch section of tin-can sheet metal. Fold the metal to form a ¼-inch square and solder it to the head of the machine screw as shown in the drawing. If you are using the aluminum foil leaves, solder a short length of wire, bent in a ¼-inch circle, to the machine screw as shown.

Cut two leaves from the simulated gold leaf (or the aluminum foil) to size, as shown in the drawing, and mount them on the end of the machine screw. The gold leaf is mounted by carefully bending the ends of the metal tab over the sides of the leaves. Do not apply too much pressure, or the gold leaf will be torn. The aluminum foil is installed with small holes punched in the top center of each leaf. Bend the wire end on the machine screw so that the leaves will swing freely, but will not fall off.

Mount a metal knob on the outside end of the machine screw as shown in the drawing and use nuts, if necessary, to keep the entire assembly from moving out of the bottle stopper.

Materials List for Gold-Leaf Electroscope

Simulated gold leaf or household aluminum foil (See text.)	Threaded rod or machine screw
Metal knob	Tin-can metal strip or solid wire (See text.)
Glass flask or bottle	Plastic or cork stopper for bottle

Electroscope Operation and Experiments

The electroscope is a sensitive indicator of electrostatic energy. It should not be used where there is a high level of electrostatic potential (such as the Van de Graaff generator), because the foil leaves may be torn apart and damaged. If possible, use a "proof plane" for experiments involving high-energy levels. The proof plane is made of a metal disc or knob attached to the end of an insulating handle. (Size is not critical—a copper penny cemented to the end of a plastic rod will work.) Touch the metal disc of the proof plane to the high-level electrostatic source; a portion of the charge will be picked up by the proof plane. To transfer the charge, carry the proof plane to the electroscope and touch the metal disc to the electroscope metal knob. The proof plane is also useful in experiments that require the electroscope to be tilted or inserted into a cavity. Several connections of a proof plane to a electrostatic source, followed by transfers to the electroscope, will show an additive effect; the electroscope leaves will spread farther apart with each transfer of energy.

The electroscope will show an indication with either negative or positive electrical charges. The distance between the leaves will show a rough indication of the strength of the charge (the farther apart they are, the larger the charge). A charge of one polarity can be neutralized by the addition of another charge of the opposite polarity.

As shown in Fig. 3-9, an electroscope can be used to demonstrate electrostatic induction by charging with one polarity, grounding, then removing the ground before the electrically charged object is removed. This can be done with either polarity of electrical charge.

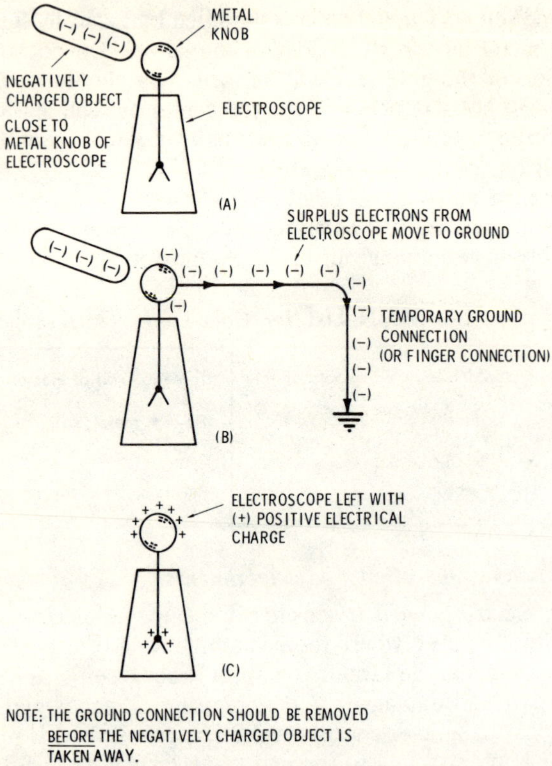

Fig. 3-9. Charging an electroscope by induction.

Electroscope Theory

As shown in the drawing of Fig. 3-10, the foil leaves of an uncharged electroscope are normally in a vertical (downward) position and close to each other.

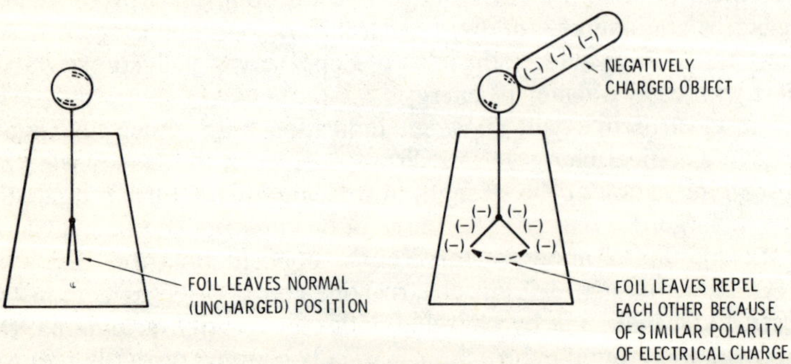

Fig. 3-10. Gold-leaf electroscope theory.

When an electrostatically charged object is touched to the metal knob on top of the electroscope, the electrical charges are conducted to the metal foil leaves. Since both of the leaves are connected together at the top, the leaves will both receive the same amount of charge and at the same polarity. Because leaves have the same polarity, they will repel each other and move apart. The larger the amount of electrical energy of the charged object, the greater the distance the foil leaves will move apart. This action will occur for either a negatively charged object, as shown in Fig. 3-11, or for a positively charged object.

Fig. 3-11. Testing the gold-leaf electroscope with a charged plastic rod.

SHIELDED ELECTROSCOPE

Construction of the Shielded Electroscope

As shown in Fig. 3-12, the shielded electroscope is built in a metal "Band-Aid" box. Any similar size metal box (approximately 3 inches wide by 3½ inches high by 1½ inches deep) can be used (Fig. 3-13).

Begin construction by carefully cutting away and removing the hinge for the box top. Also, scrape away the paint around the top rim. If necessary, bend the box top sides to provide a secure fit, but for easy removal.

Cut a 1½-inch-diameter hole in the center of each wide side of the box and install sections of clear plastic sheet on the inside as shown. Cut a 1-inch hole in the center of the lid and install a section of plastic with a center hole drilled to fit the 1¾-inch-long machine screw.

Cut tin-can metal section C as shown in Fig. 3-13 and solder the

Fig. 3-12. Shielded electroscope project.

end of the section to the head of the machine screw. Cut a ¼-inch-wide by 1-inch-long section of simulated gold leaf or aluminum foil and install it on the C section by bending over the ⅛-inch tabs to hold the foil section in place.

Mount the machine-screw assembly with nuts in the top plastic section and install a metal knob on the screw end. Adjust the length of the machine screw so that the head of the machine screw (when the lid is installed) is just below the top of the side holes.

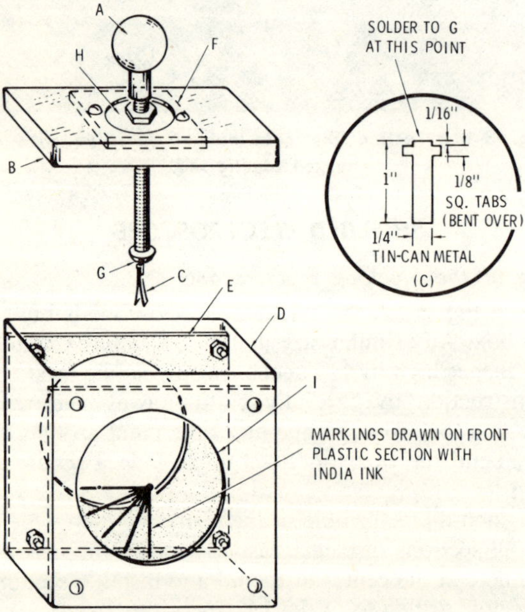

Fig. 3-13. Plans for shielded electroscope.

Install the lid on the box and make sure that the foil swings freely when the box is tilted to one side. We painted the box and drew markings on one of the side-hole plastic sections, as shown in the drawing.

NOTE: Operation and theory is the same as the gold leaf electroscope (see Fig. 3-10).

Materials List for Shielded Electroscope

A—Metal knob B—Top lid of E C—Tin-can steel section approximately 1/4" wide × 1" long with 1/4" wide × 1" long simulated gold leaf or aluminum foil clamped with a tab of tin-can steel at the top of the D section.	D—Metal box (with lid) approximately 3" wide × 3 1/2" high × 1 1/2" deep metal "Band-Aid" box or equivalent E—Clear sheet-plastic sections to fit the front and back of E F—Plastic section to fit H G—1 3/4" long, 6-32 machine screw

TWIN-ELECTRODE FOIL ELECTROSCOPE

Construction of the Twin-Electrode Foil Electroscope

This electroscope is built in a clear plastic jar, the type used for household food storage (Fig. 3-14). The size is not critical; we used a 4-inch-diameter by 3-inch-high clear plastic jar. For best results, follow the parts layout, as shown in Fig. 3-15. Note that the jar is used upside down, with the lid on the bottom. Either a screw-on or press-on lid can be used as long as it is flat and does not have protrusions.

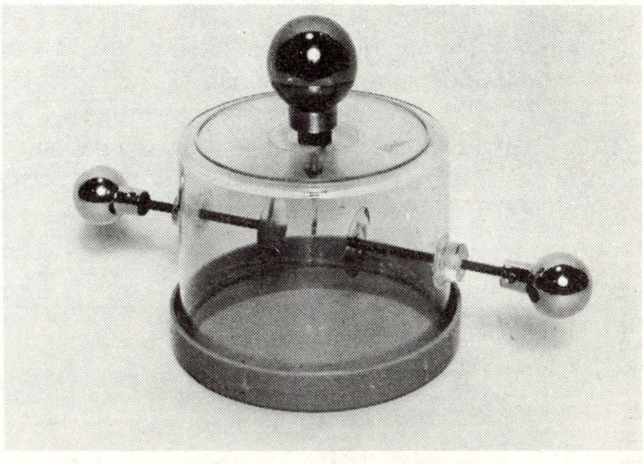

Fig. 3-14. Twin-electrode foil electroscope.

Cut two holes (located on opposite sides) approximately 1½ inch down from the bottom of the jar and cement ½-inch plastic bushings with threaded holes to accept 2½-inch machine screws, as shown in the drawing.

Cut two 1-inch diameter discs from tin-can metal and solder them to the heads of the 2½-inch machine screws. Then, install the machine screws in the threaded plastic bushings and mount metal knobs on the screw ends.

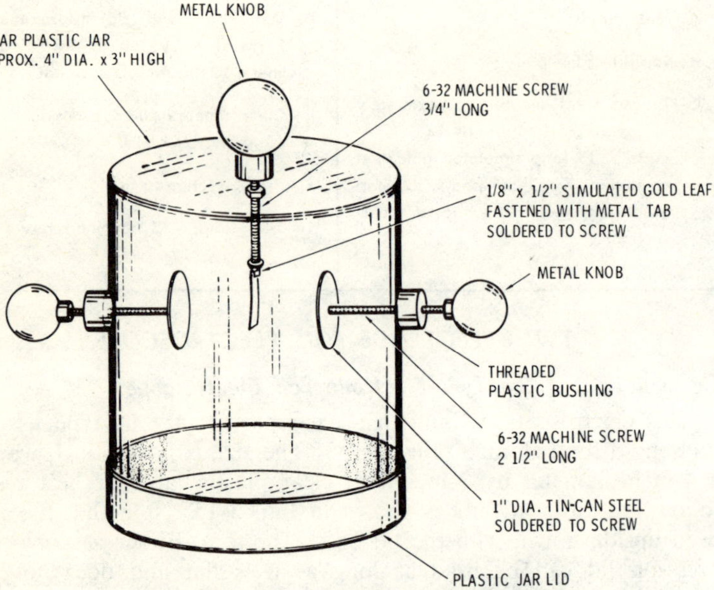

Fig. 3-15. Plans for twin-electrode foil electroscope.

Mount a ⅛- by ½-inch section of simulated gold leaf with a ¼-inch tin-can metal tab soldered to the head of a ¾-inch machine screw, as detailed in Fig. 3-7 of the gold-leaf electroscope project. Or, you can use an ¼- by ¾-inch aluminum foil section mounted with a wire bent in a ¼-inch circle and soldered to the machine screw head. (As also detailed in the gold-leaf electroscope project.)

Materials List for Twin-Electrode Foil Electroscope

Simulated gold leaf or aluminum foil (See text.)	Threaded rods and nuts
Metal knobs	Tin-can metal
Plastic jar	Plastic rod (for bushing)

Install the machine screw in the center hole of the jar bottom with nuts and a metal knob on the screw end. Adjust the length of the screw so that the top of the foil leaf is even with the tops of the metal discs. You can use small cardboard washers between the screw mounting nuts and the jar bottom to avoid fracturing the plastic.

Twin-Electrode Foil Electroscope Theory and Operation

This type of electroscope can be used as a polarity indicator for unknown electrostatic objects. For example, it can be used to determine the polarity of charges of different types of plastics and other materials.

A dc power supply, approximately 300 volts, is connected to the parallel metal discs as shown in Fig. 3-16 (or to a charged capacitor

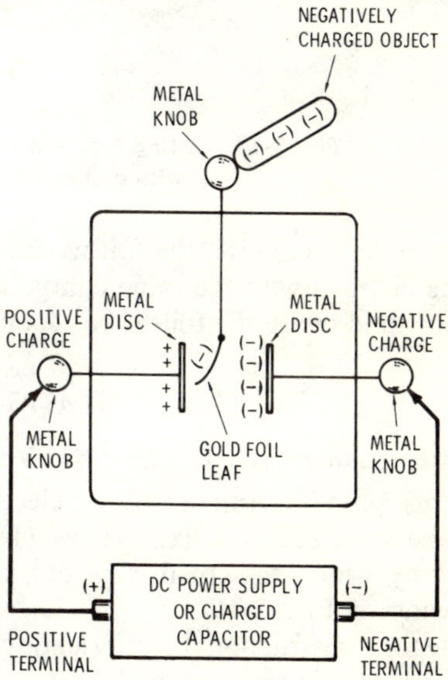

Fig. 3-16. Twin-electrode foil electroscope theory.

or Leyden jar). When a negatively charged object is connected to the gold foil metal knob (on top), the surplus electrons on the gold foil will be attracted toward the positively charged metal disc, thus moving the foil in the direction of the positive metal disc (Fig. 3-17).

If a positively charged object is connected to the foil metal knob, the positive charge on the foil will cause the foil to be attracted toward the negatively charged metal disc.

The metal discs on both sides of the foil leaf should be adjusted for both the expected electrostatic potential of the unknown object and the voltage of the dc power supply. The greater the voltage of the power supply, the wider the spacing should be. The spacing should be wide enough so that the foil leaf does not touch the metal disc on

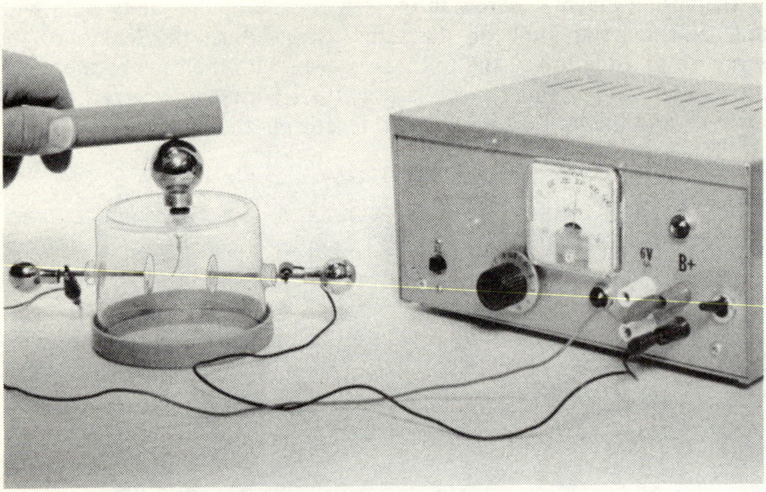

Fig. 3-17. Testing the twin-electrode foil electroscope with a charged plastic rod.

either side; otherwise, the foil would lose its electrical charge and become charged with the same charge polarity as that of the metal disc. This would cause the foil to be repelled by the disc.

ELECTROSTATIC VOLTMETER

Construction of the Electrostatic Voltmeter

This simplified model of an electrostatic voltmeter has a moving double vane and two fixed vanes (Fig. 3-18). It is built on ⅛-inch-thick by 4-inch-wide by 6-inch-long plastic baseboard, labeled C in the drawing of Fig. 3-19.

Begin construction by fabricating bracket G from 1- by ½- by ¼-inch plastic sections, as shown in the components drawing of Fig. 3-20. After cementing the bracket together, cement a 1- by ¼- by ¼-inch section to one side under the bracket to allow clearance for rubber band F. Set the bracket aside and cut out 2¾- by 3¾-inch cardboard scale A. Mount the cardboard scale with four ¾ inch diameter by ¾-inch-high plastic B sections on the baseboard, as shown in the drawing.

Cement the bracket to the baseboard about 1½ inches away from the edge of the scale and centered as shown in Fig. 3-19. Cut out moving vane E from tin-can metal, as shown in the drawing of Fig. 3-20. Select a rubber band that will fit around the bracket. The rubber band should be about 1/16 inch wide and not too tight. If necessary, cement an additional plastic section on the back of the bracket to fit the

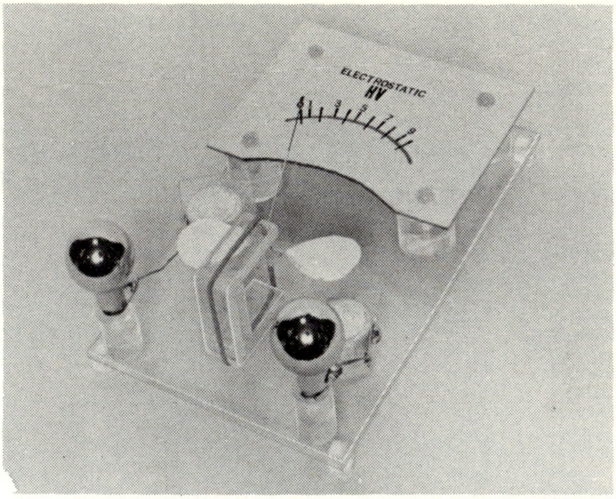

Fig. 3-18. The electrostatic voltmeter project.

rubber band. Mount the moving vane on the rubber band by bending the tabs over the rubber, as shown in the drawing. Then twist the vanes in opposite directions so that they are parallel to the baseboard.

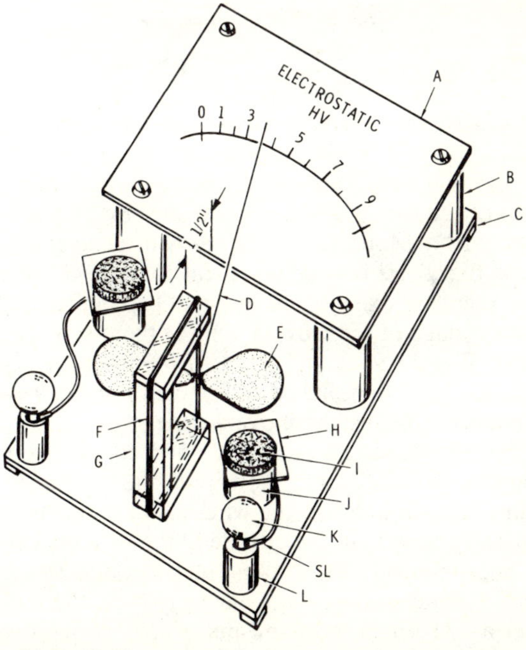

Fig. 3-19. Electrostatic voltmeter assembly drawing.

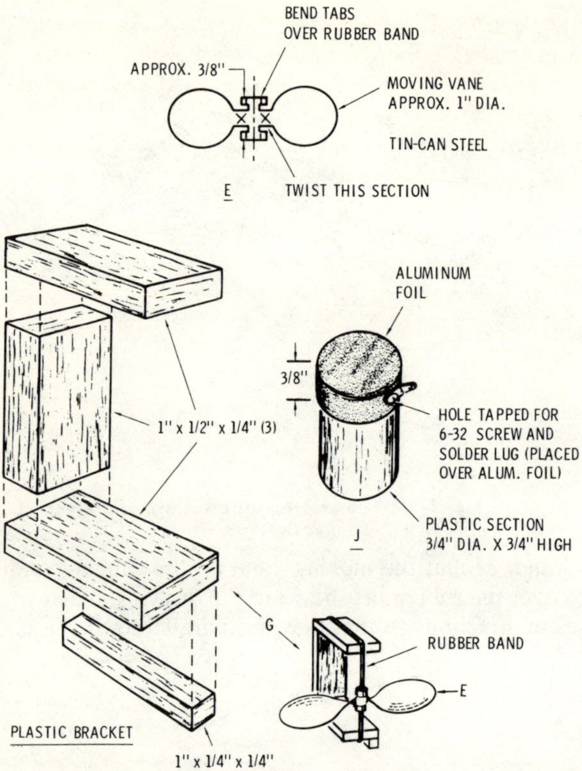

Fig. 3-20. Electrostatic voltmeter components drawing.

Fabricate the two fixed vanes, J, by cementing aluminum foil over about ⅜ inch of the top of a ¾ inch diameter by ¾ inch high plastic section (Fig. 3-20). Mount a solder lug with a machine screw in a tapped hole so that there is good electrical contact with the aluminum foil. Cement the two fixed vanes on the baseboard approximately ⅛ inch from the edge of the moving vane in the positions as shown in the drawing of Fig. 3-19.

Cut out section H of thin plastic and cement it over the top and side of the aluminum foil facing the moving vane.

Stretch a length of No. 28 enameled solid copper wire as straight as possible and solder one end of the wire to the top moving vane tab. You can clamp the tab between the wire end and the rubber band instead of soldering the wire to the tab. Bend the wire over the scale and measure off approximately 2½ inches; then cut off the excess wire to make pointer D. If desired, you can carefully bend the free end of the wire in about a 1/16-inch loop and put a drop of paint on it (as in our model) so it will be more visible.

Cut two plastic L sections approximately ½ inch in diameter by ¾ inch high and drill center holes. Thread the top ends for machine screws to fit metal knobs, K, and solder lugs, SL. Cement the assemblies to the corners of the baseboard, as shown in the drawing of Fig. 3-19. Connect the solder lugs on the L and K terminals to the solder lugs of the fixed vanes with short lengths of solid wire that are bent away from the path of the moving vane.

Draw a scale on A (our model was numbered 0 to 9), and adjust the pointer and rubber band as necessary to position the pointer at scale 0. Adjust the moving vane and rubber band vertically so that the moving vane is about ⅛ inch above the top of the fixed vanes.

Materials List for Electrostatic Voltmeter

A—Cardboard scale 2¾″ × 3¾″	F—Rubber band (See text.)
B—Plastic section ¾″ diameter × ¾″ high	G—Plastic bracket (See Fig. 3-20.)
C—Plastic baseboard 4″ × 6″ × ⅛″	H—Thin sheet plastic
D—Pointer: No. 28 enameled copper wire approximately 3″ long	I—Aluminum foil
E—Moving vanes of tin-can steel (See Fig. 3-20.)	J—Fixed vane (See Fig. 3-20.)
	K—Metal knob
	L—Plastic section, ½″ diameter × ¾″ high

Electrostatic Voltmeter Operation and Experiments

One of the foil electrodes is normally connected to ground, and the other electrode is used for connection to the electrically charged object. The electrodes are interchangeable; they have no polarity. Either one can be used for the ground or charge connection. Metal knobs are used for both of the electrodes, but, for convenience, one of the knobs can be removed and replaced with a binding post for easier connections to ground leads. The meter can be used uncalibrated for relative strengths of electrostatic charged objects, or it can be compared to known high voltages and calibrated accordingly.

You can easily test the voltmeter by rubbing a plastic rod with a woolen cloth and connecting the charged end to one of the meter electrodes, or by using the electrophorus as in Fig. 3-21 (the other electrode should be connected to an external ground). The meter pointer should show an indication. A greater meter indication can be seen by holding the meter by one of the knobs (you are then the ground connection) and placing the other knob electrode near an operating Van de Graaff Generator.

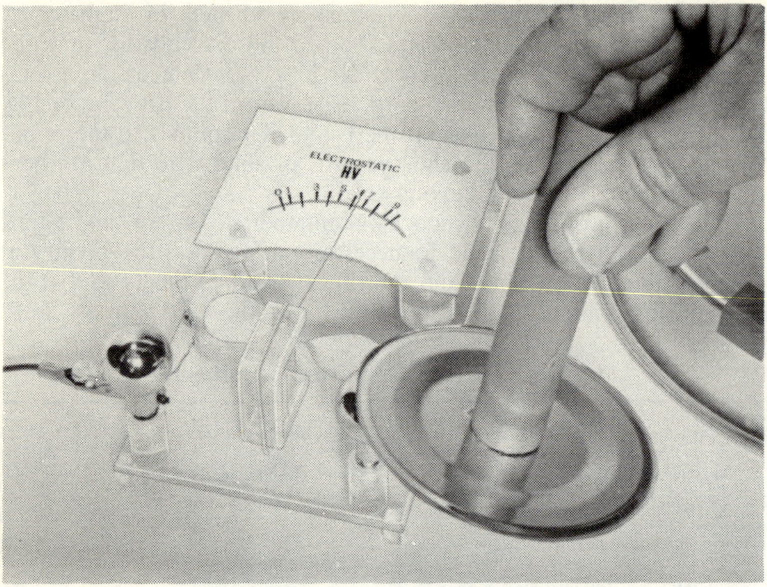

Fig. 3-21. Testing the electrostatic voltmeter project with the electrophorus charge carrier.

Try using the voltmeter with different types of electrostatically charged materials to check the amount of electrical charge for each type. After each measurement, discharge the meter by connecting a lead between the knob electrodes. If necessary, reset the meter to zero by repositioning the rubber band.

Electrostatic Voltmeter Theory

This voltmeter is similar to the versorium in principle. One of the foil electrodes is connected to ground, and the other foil electrode is connected to the charged object, as shown in Fig. 3-22. The voltmeter will operate on either negatively or positively charged objects.

When a positively charged object is connected to the foil electrode, as shown in the drawing, surplus electrons of the moving metal vane try to move closer to the foil electrode, attracting the vane toward the foil. At the same time, the positive charges on the other end of the moving vane (formed by the movement of the surplus electrons) are attracted toward the foil electrode connected to ground. This electrode has a negative charge on it, formed by the electrons from the ground connection that are attracted by the positive charge on the moving vane end.

The moving vane rotates against the slight tension of the rubber-band suspension and moves the attached pointer over the scale. The

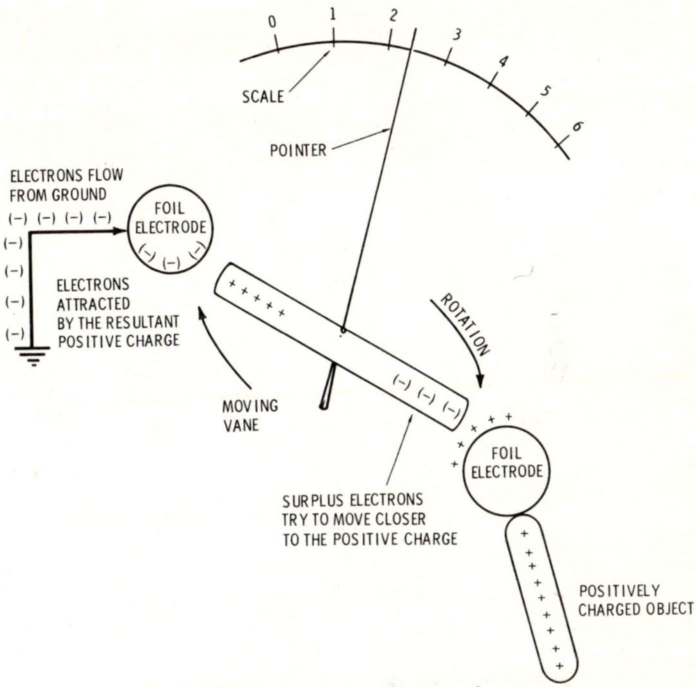

Fig. 3-22. Electrostatic voltmeter theory.

amount of movement of the vane and pointer is dependent upon the amount of electrostatic potential of the charged object; the higher the electrostatic potential, the closer the moving vane moves to the foil electrodes, moving the pointer over more of the scale area.

When the charged object is removed, the moving vane will be pulled by the rubber band suspension back to approximately normal (uncharged), or zero, position. It may be necessary to connect both electrodes to ground to discharge any residual electrostatic energy left in the voltmeter. When a negatively charged object is connected, the electrostatic attractions are reversed, but the meter operation is the same.

CHAPTER 4

From Leyden Jars to Capacitors

The Leyden jar is the early type of device used to store electrical energy in electrostatic form. This device has gone through many changes in shape and names. The first name, "Leyden jar," was in honor of the University of Leiden, where van Musschenbroek made his famous experiment with a glass bottle full of water charged from an electrostatic generator. A second name, "condenser," goes back to the theory of early experimenters who thought that electricity was an "electrical fluid" which could be "condensed." The term condenser has been used up to modern times, and it is now superseded by "capacitor." This device for storing electrical energy still operates in the same manner, even though it evolved from bottles of water, to jars full of metal shot, and then to metal foils and dielectrics.

The early experimenters employed small children to dangle from silken threads with their feet touching the spinning, hand-rubbed sulfur or glass globes of the electrostatic generators. The electrical charges for experiments were taken from their bodies. Instead of children, later experimenters used metal tubes (even gun barrels), suspended from silk cords. These metal tubes were connected to the electrostatic generators by metal chains dangling close to the surface of the rotating globes and were called "prime conductors or collectors." The tubes could store a small amount of electrostatic energy on their charged metal surfaces. Metal chains on the free end of the prime conductors were connected to electrostatic experiments.

It was such a metal chain that Cuneus, a student of van Musschenbroek, tried to remove while charging the first "Leyden jar" and he received a severe electrical shock. Therefore, observe proper precautions while performing experiments with Leyden jars or capacitors. Do not touch *both terminals* of a charged Leyden jar or capacitor at the same time. Always discharge a Leyden jar or capacitor by shorting out the terminals with a lead before putting them away in storage, or at the end of each experiment. Always handle all Leyden jars and capacitors as if they are electrically charged.

WATER BOTTLE LEYDEN JAR

Water Bottle Leyden Jar Operation and Experiments

The original Leyden jar discovery was actually a glass of water with a chain from an electrostatic generator dangling in it. Our version uses a solid copper wire held in a corked bottle, as shown in Fig. 4-1.

The dimensions of the bottle are not important, any size bottle will do (our model used a bottle 2 inches in diameter by 3¾ inches high.)

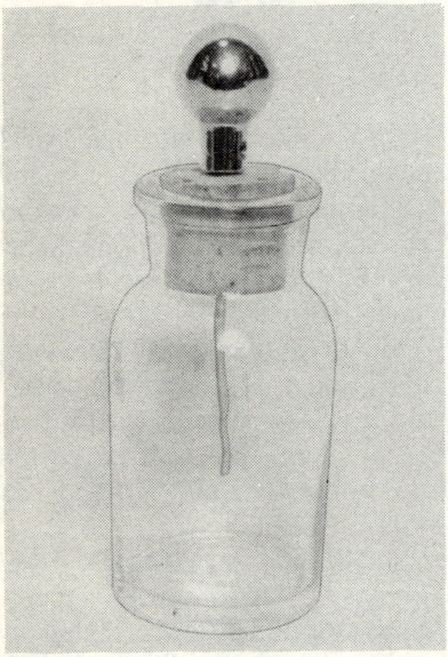

Fig. 4-1. Water bottle Leyden jar project.

The copper wire should extend into the bottle, from the cork, about three-quarters of the height of the bottle. Fill the bottle to a level close to the cork, leaving an air space (Fig. 4-2).

The water bottle Leyden jar operates the same as the Leyden jar with inner and outer foils (see Fig. 4-6). However, water takes the place of the inner foil, and your hand takes the place of the outer foil (like the type of Leyden jar the early experimenters used).

Hold the bottle in one hand and charge the bottle by holding its metal knob near a high-voltage electrostatic generator (Fig. 4-3). Then you should be able to light a neon lamp (NE-2 or equivalent) by connecting one lamp lead to ground and the other lead to the metal knob (while holding the outside of the bottle). Do not touch the metal knob with your hand while holding the outside of the bottle. If you do

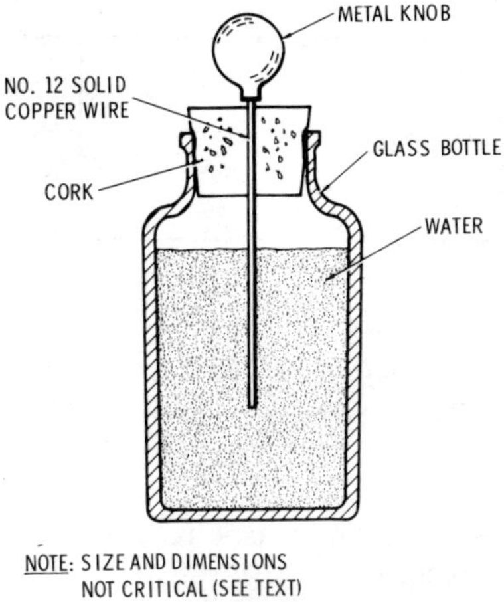

Fig. 4-2. Drawing of water bottle Leyden jar.

touch it, the electrostatic energy will discharge through you, causing you to receive a shock like the original discoverer.

Wrap a section of aluminum foil around the bottle and connect the foil to ground. Repeat the previous experiment, and you should find that the neon lamp will have a much brighter flash. This is because the aluminum foil is a better conductor than your body and has a direct connection to ground.

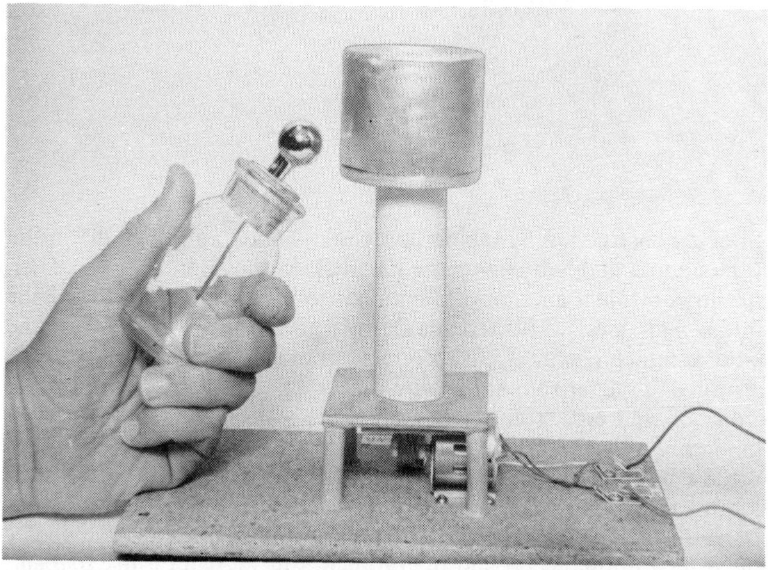

Fig. 4-3. Charging the water bottle Leyden jar from a Van de Graaff generator.

Materials List for Water Bottle Leyden Jar

Metal knob	Glass bottle
Solid copper wire	Water
Cork	

LEYDEN JAR

Construction of the Leyden Jar

The Leyden jar is built in a clear plastic jar approximately 3 inches in diameter by 3½ inches high with a plastic lid (Fig. 4-4). The size is not critical, and any size jar can be used. A glass jar can also be used, but the plastic jar has better electrical qualities (lower electrical leakage).

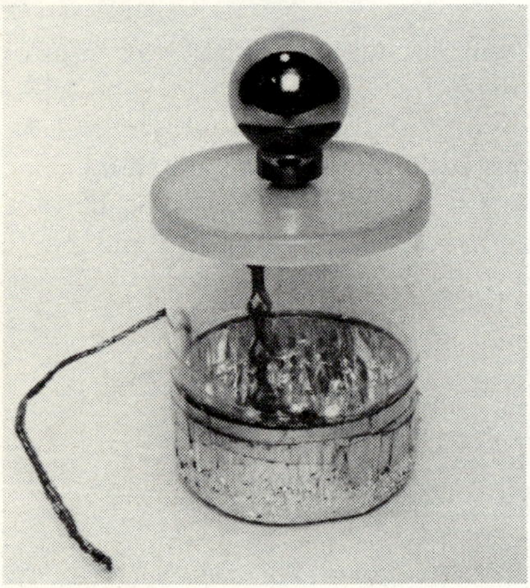

Fig. 4-4. Leyden jar project.

Begin construction by taping household aluminum foil to the inside of the bottom and halfway up the jar sides, as shown in Fig. 4-5. Then tape an equivalent amount of aluminum foil around the outside of the bottom and sides, with the ends parallel to the inside foil. Insert the end of a 6-inch length of bare, copper, stranded-wire between the outside foil and the jar so that the wire makes a good electrical connection to the foil and tape it in place.

Connect a length of light metal chain to a machine screw. The chain should be long enough to reach the bottom of the inside foil (draped loosely) from the jar top, as shown in the drawing. We used 8 inches of chain in our model. Install the machine screw with nuts in a center hole drilled in the jar lid and then mount a metal ball on the free end of the screw outside of the jar lid.

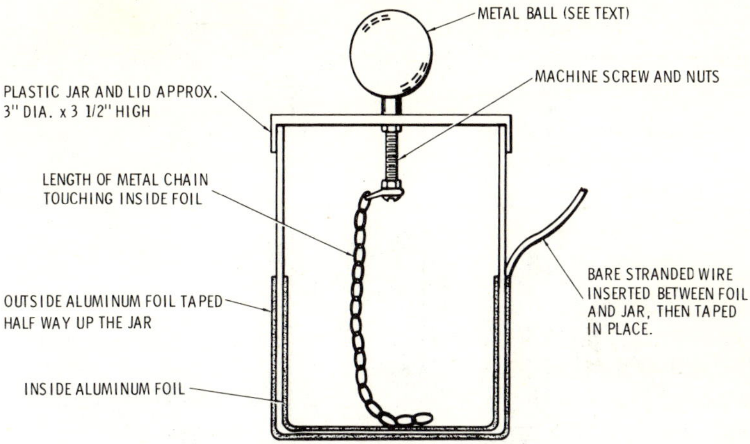

Fig. 4-5. Leyden jar project drawing.

Install the lid on the jar and tilt the jar to make sure that the metal chain touches the foil for a good electrical connection.

Materials Lists for Leyden Jar

Metal knob or metal ball (See text.)	Aluminum foil
Machine screw and nuts	Stranded, bare, copper wire
Plastic jar (with plastic lid)	Plastic tape
Metal chain	

Leyden Jar Theory

The Leyden jar is an early form of capacitor (or condenser). When an electrostatic or dc voltage source is connected to the Leyden jar, as shown in Fig. 4-6, the inside and outside metal foils produce an electrostatic field between them.

As shown in Fig. 4-6, a positive voltage connected to the inside foil has a positive electrostatic field, and the negative voltage on the outside field has a negative electrostatic field. These fields produce an internal stress on the plastic container between the foils and force negative charges to appear on the inside surface of the plastic (adjacent to the positively charged inner foil). The negatively charged outside foil produces positive charges on the outer surface of the plastic.

The higher the electrostatic voltage applied to the Leyden jar, the larger the internal stresses and the greater the amount of positive and negative charges produced on the plastic surface. The plastic between the two foils is called the dielectric. If the voltage applied to the Leyden

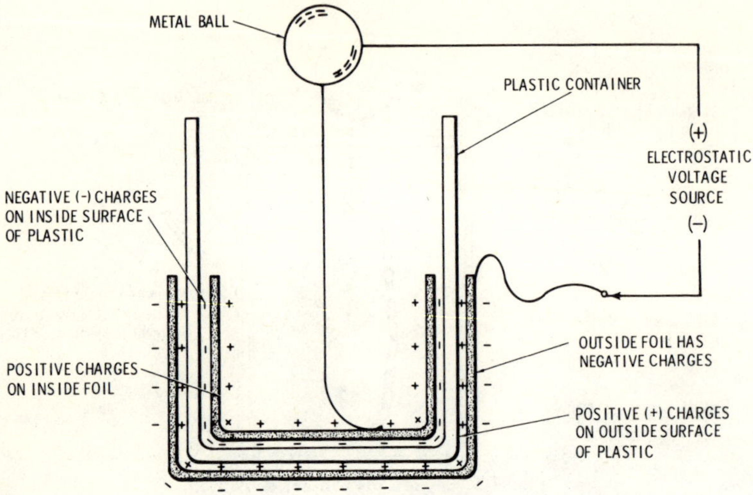

Fig. 4-6. Leyden jar theory.

jar (or capacitor) is high enough, the stresses of the dielectric may become too great, and the plastic will physically break down and short out internally.

The action of the Leyden jar is due to induction and may be considered similar to that of the electrophorus, except that the electrophorus plastic sheet (or dielectric) is mechanically rubbed with a cloth to produce the electrostatic charges.

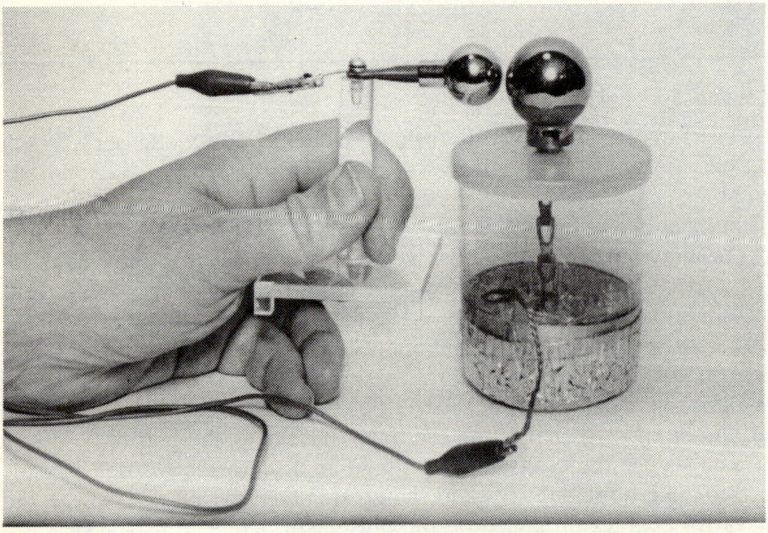

Fig. 4-7. Discharging the Leyden jar project with a spark-gap assembly.

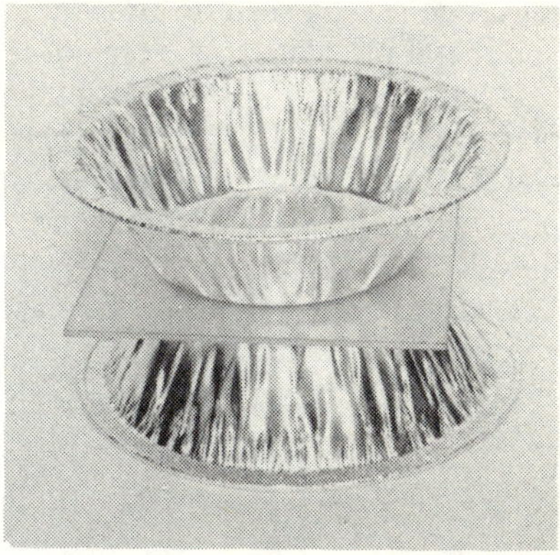

Fig. 4-8. Franklin's Leyden jar dielectric experiment. (Modern version.)

When the voltage source is removed from the Leyden jar (or capacitor), the electrical charges on the plastic dielectric may remain for a long time, keeping the electrical energy in storage. When the Leyden jar terminals are connected together (Fig. 4-7), the electrostatic energy is given up by the dielectric to the foils and to the external circuit.

FRANKLIN'S LEYDEN JAR DIELECTRIC EXPERIMENT

Franklin's original experiment used glass jars and glass sheets for the dielectric, and sheet lead for the electrodes. We are using thin sheet

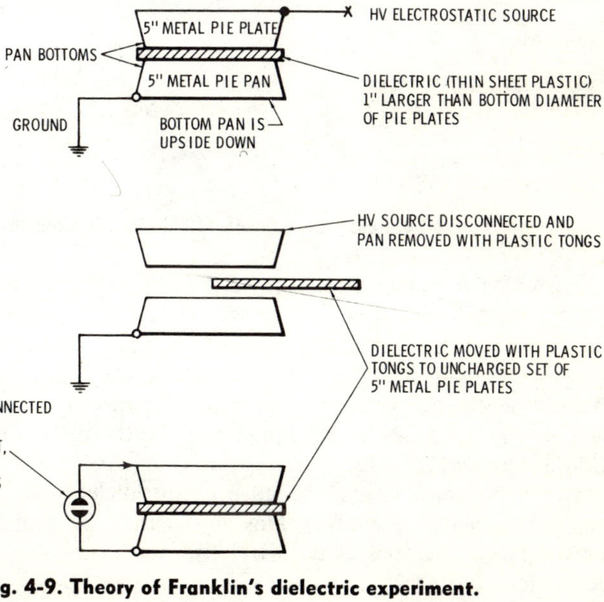

Fig. 4-9. Theory of Franklin's dielectric experiment.

plastic (the type used for notebook sheet protectors) for the dielectric, and metal pie plates for electrodes (Fig. 4-8).

As shown in Fig. 4-9, ground the bottom plate and cut the thin plastic sheet to a size about 1 inch larger than the diameter of the pie plates. Insert the plastic sheet between the two pie plates (the bottom plate is upside down) so that the two plate bottoms are separated by the plastic.

Charge the top plate of the capacitor assembly by placing a high-voltage electrostatic generator (such as the Van de Graaff generator) close enough to the rim of the plate so that sparks will jump from the generator.

Fig. 4-10. Performing Franklin's dielectric experiment.

Turn off the generator and move it away so that its residual electrostatic field will not affect the experiment. Pick up the top metal pie plate with a plastic clothes pin or plastic tongs and set it aside (Fig. 4-10). With plastic tongs, carefully move the sheet plastic dielectric to a pair of uncharged metal plates (as shown in Fig. 4-9). Then connect a neon lamp (NE-2 or equivalent) to the plates. The neon lamp should momentarily light, indicating that the uncharged capacitor assembly became charged because of the dielectric moved from the charged capacitor assembly. This shows that the charge in a capacitor (or Leyden jar) is in the dielectric, not in the metal foil or plate electrodes.

Fig. 4-11. Paper capacitor experiment.

Materials List for Dielectric Experiment

Metal pie plates	NE-2 neon lamp
Thin sheet plastic	Plastic tongs

PAPER CAPACITOR EXPERIMENT

Try making your own capacitors as the early experimenters did. As shown in Fig. 4-11, materials are ordinary household aluminum foil and waxed paper. Construction is easy and lends itself to many variations in fabrication. Thin plastic sheet can be used in place of

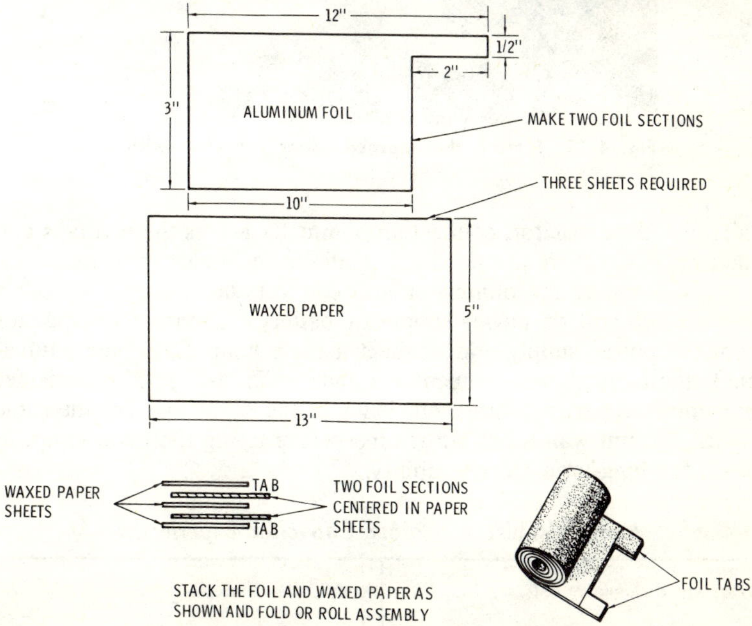

Fig. 4-12. Paper capacitor project plans.

the waxed paper, and the dimensions can be varied for different values of capacitance (larger area of foil—larger capacitance). Make sure that there are no holes in the waxed paper, because the foils may short circuit and ruin the capacitor.

Begin construction by cutting the foils and waxed paper to the approximate sizes shown in Fig. 4-12. Then place the foil sections on the center of the paper sheets, positioned so that the foil tabs are in reverse positions on far sides of the paper. This is necessary to prevent any accidental contact between the two. Make sure that there is at least 1 inch of paper all the way around each foil section (except for the tabs) for best insulation between the foils.

After stacking the foil sections as shown in the drawing, fold or roll the capacitor assembly as tightly as possible and place a rubber band or plastic tape around the capacitor to keep the turns in place.

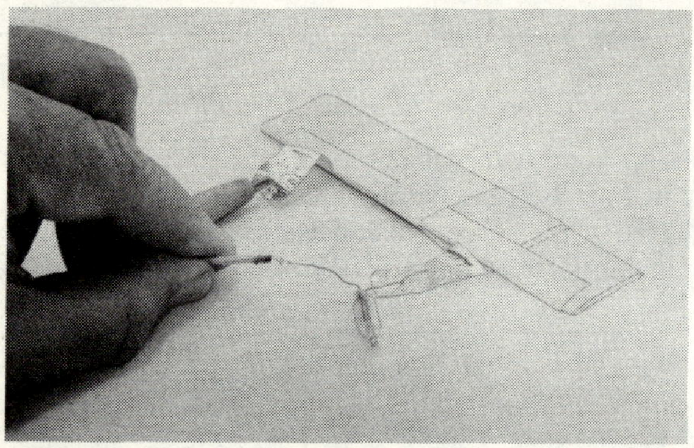

Fig. 4-13. Testing the charged paper capacitor with a neon lamp.

To test the capacitor, connect an ohmmeter across the two tabs and make sure that there are no shorts (infinite resistance measurement). Then, disconnect the ohmmeter and connect the capacitor across a 100- to 300-volt dc power supply or battery. Remove the capacitor from the power supply and connect a neon lamp (NE-2 or equivalent); the lamp should momentarily light (Fig. 4-13). This indicates momentary electrical storage efficiency of the capacitor. Try the same experiment but wait a period of time before trying the neon lamp, to check for long-term storage ability.

Materials List for Paper Capacitor Experiment

Household aluminum foil	NE-2 neon lamp
Waxed paper	

CHAPTER 5

Electrostatic Motors

The first electrostatic motor was invented by Benjamin Franklin in 1748. He developed his motor during his research on the theory and operation of the Leyden jar. His motor was a very simple type that used two oppositely charged Leyden jars for electrical power.

After Franklin's time, very little was done on the design of electrostatic motors until 1870, when the German physicist J. C. Poggendorf, built a motor with a glass disc and two electrodes that sprayed electrical charges on the disc surface to make it rotate. This motor was designed to be used with a high-voltage electrostatic generator such as the Wimshurst machine.

In recent years there has been a renewed interest in the design of electrostatic motors. In articles in the April 1971 and May 1971 issues of *Popular Science* magazine (which should be available at your local library), experimental electrostatic motors are described that have been built at the University of West Virginia by Dr. Oleg Jefimenko.

In this chapter, basic types of electrostatic motors (Fig. 5-1) are included as construction projects. The field of design of electrostatic motors is wide open. After constructing the motors described, try designing your own.

FRANKLIN MOTOR

Construction of the Franklin Motor

Our version of Franklin's motor is constructed as shown in photo Fig. 5-2. The motor is built on a 2- by 3- by 3/16-inch plastic baseboard with ¼ inch square corner feet (Fig. 5-3).

Begin construction by cementing a 1¾-inch-high by ½-inch-diameter plastic rod in the center of the baseboard. Mount a long needle in the center of the rod (the needle on our model is approximately 2

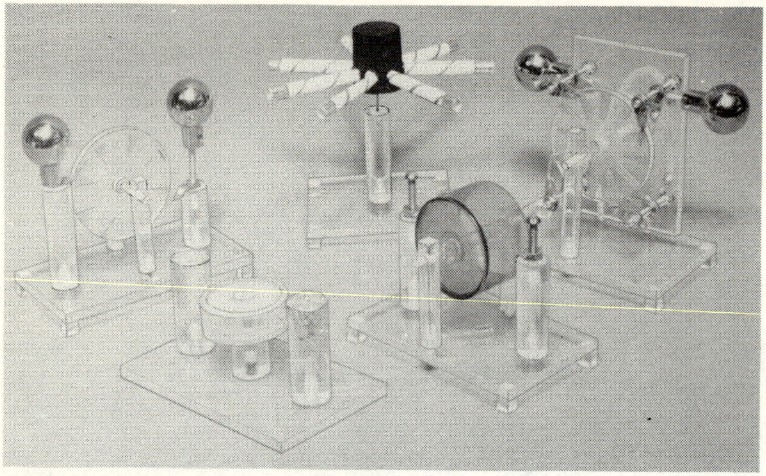

Fig. 5-1. Electrostatic motor projects included in this chapter: (left to right) two-brush disc motor, plastic-disc motor, Franklin motor, plastic-drum motor, four-brush disc motor.

inches long). An easy way to mount the needle is to heat the blunt end (the eye section) and press it into the plastic. Make sure that the needle is exactly vertical.

Select a plastic bottle cap that is approximately 1-inch long with a 1-inch diameter at the base. Carefully drill seven holes, large enough to hold drinking straws, evenly spaced around the base of the plastic bottle top as shown in the drawing.

Cut seven 2-inch sections of either plastic or paper drinking straws and cement a ¼-inch ring of household aluminum foil on one end of each straw section. Cement the other end of each straw section into the bottle cap, located as shown in the drawing, and balance the cap assembly on the needle tip.

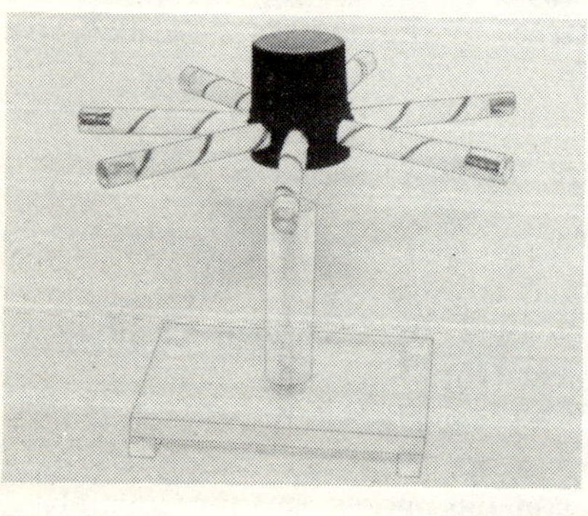

Fig. 5-2. Franklin motor project.

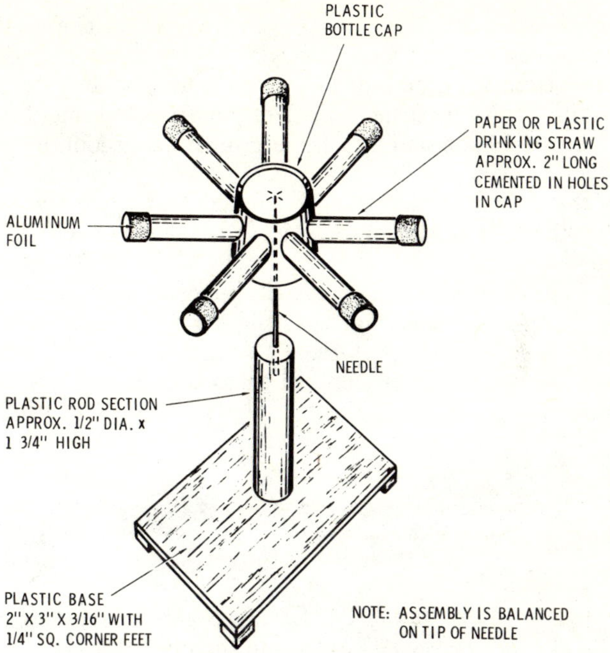

Fig. 5-3. Franklin motor project plans.

Drill or center punch a small indentation in the inside of the cap center to hold the assembly balanced (do not drill all the way through the cap). If necessary, change the straw lengths slightly for best balance.

Materials List for Franklin Motor

Plastic rod	Paper or plastic drinking straws
Plastic section for base (See text.)	Plastic bottle cap
Needle	Aluminum foil

Franklin Motor Operation and Experiments

The Franklin electrostatic motor requires a minimum of two charged Leyden jars for operation. These jars should be charged from an electrostatic source. The only requirement is that the Leyden jars should be charged to opposite polarities (one jar negatively charged and the other jar positively charged). Since many electrostatic generators have only one polarity output (our model of the Van de Graaff has a positive electrostatic charge output), the following method can be used to obtain an opposite charge output.

Ground the metal knob of the Leyden jar and charge the jar by holding the outside foil close enough to the generator so that sparks will jump from the generator to the foil. Then, holding the jar by the plastic rim (or with plastic tongs) disconnect the ground connection from the metal knob and attach the ground connection to the outside foil covering. The polarity of the metal knob will then be opposite to that of the charging electrostatic generator. Make sure that you do not touch both the Leyden jar outside foil and the metal knob at the same time.

Place the charged Leyden jars on opposite sides of the Franklin motor, close to the ends, and give the motor vanes a push in either direction to get the motor started. The operation of this motor takes considerable patience in placing the charged Leyden jars in the proper positions.

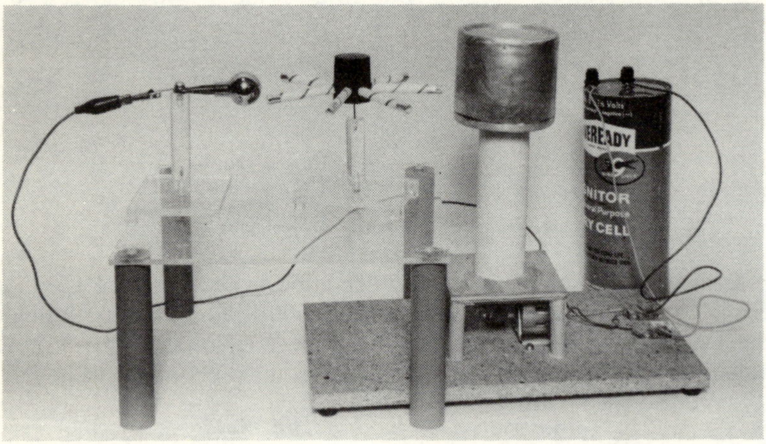

Fig. 5-4. Testing the Franklin motor project with a Van de Graaff generator and a grounded spark-gap assembly.

You can experiment with different positions of the Leyden jars by adding more jars, grounded electrodes, and by placing the motor between a Van de Graaff generator and a grounded electrode (Fig. 5-4).

Franklin Motor Theory

The Franklin motor is positioned between two charged Leyden jars; one jar is negatively charged, and the other has a positive charge (Fig. 5-5).

The foil-covered ends of the Franklin motor receive electrical charges by corona discharge from the metal knobs of the Leyden jars and are repelled (because the foils assume the same polarity charge). Then, as the foils approach the oppositely charged Leyden jar, the

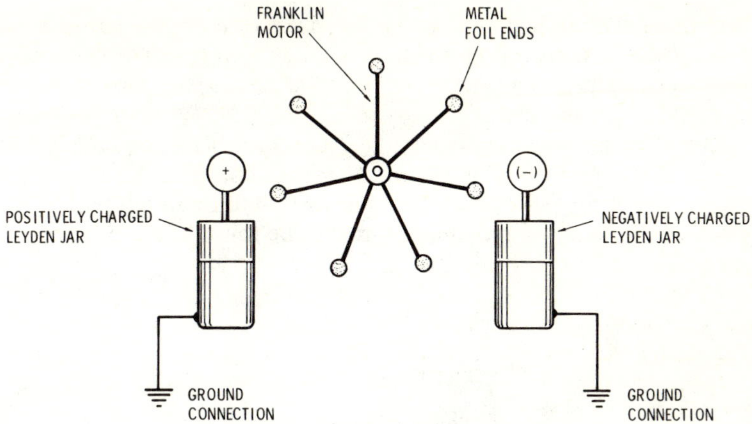

Fig. 5-5. Franklin motor theory.

foils receive a new electrical charge (the old charge becomes neutralized) and are again repelled, thus rotating the motor.

The charges from the Leyden jars are gradually dissipated as they neutralize the opposite polarity (old) electrical charges on the rotating foils. Thus, the motor will run for only a short period of time.

PLASTIC-DISC MOTOR

Construction of the Plastic-Disc Motor

This electrostatic motor is built on a 4- by 2⅜- by 3/16-inch plastic baseboard, as shown in Fig. 5-6. Cement a ⅝-inch-diameter by 1-inch-high plastic rod in the center of the baseboard, and then install a ⅜-inch section of needle or pin pointed end up in the top center of the rod (Fig. 5-7). The needle can be soldered to the head of a machine screw mounted in a tapped centerhole (as in our model), or you can heat the blunt end of the needle section and press it into the top of the plastic rod. Make sure that the needle is vertical.

The plastic bottle cap for the motor is approximately 1⅜ inch in diameter and is ½-inch-high, and balance it on the point of the needle.

Fig. 5-6. Plastic-disc motor project.

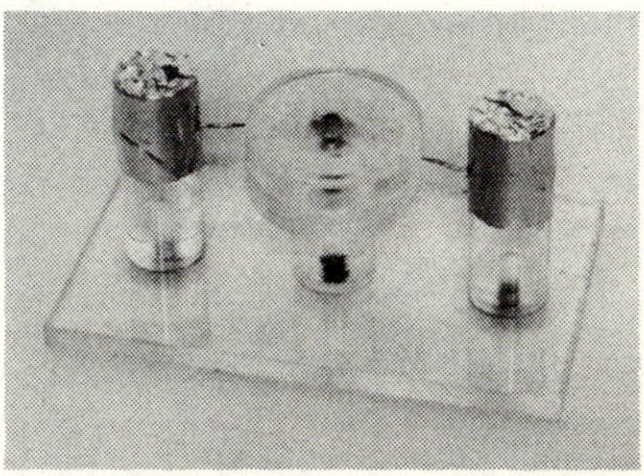

Make a mark or indentation at the balance point. If desired, you can have a better balanced disc with a bit more work by cementing a small metal cone (such as used in inexpensive pocket compasses) in the center of the bottle cap.

Cement household aluminum foil over ¾ inch of one end of two ⅝-inch-diameter by 1½-inch-high plastic rod sections. Install a ½-inch section of the pointed end of a pin in each foil covered plastic end by heating the pin and pressing it through the foil into the plastic.

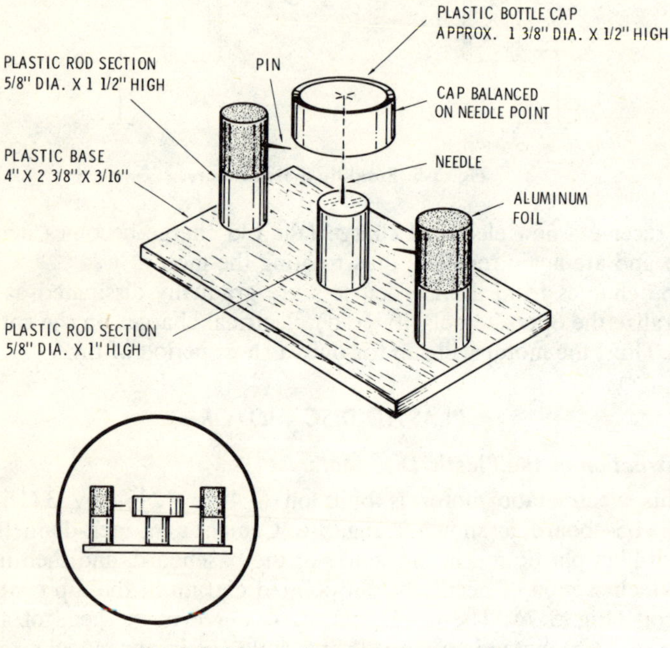

Fig. 5-7. Plastic-disc motor project plans.

Cement the plastic rod sections to the baseboard so that the pin ends just graze, but do not touch, the plastic bottle cap, as shown in the drawing. Bend the pins slightly so that they point away from each other on a horizontal plane.

Materials List for Plastic-Disc Motor

Plastic rod	Pins
Plastic section for base (See text.)	Plastic bottle cap
Needle	Aluminum foil

Plastic-Disc Motor Operation

The plastic-disc motor will operate between two oppositely charged Leyden jars; between one charged Leyden jar and a grounded electrode; and between a high-voltage electrostatic generator (such as the Van de Graaff) and a grounded electrode. This motor is very sensitive and requires little electrostatic energy to operate, if constructed properly.

A good electrical contact must be made between the motor foil electrodes and the electrostatic source. The motor can be held by one of the foil electrodes; the other foil electrode is placed in close proximity to the can of a Van de Graaff generator to operate the motor (Fig. 5-8). It can also be placed on a plastic table (see Chapter 7), so that one foil electrode is within the corona field of a Van de Graaff

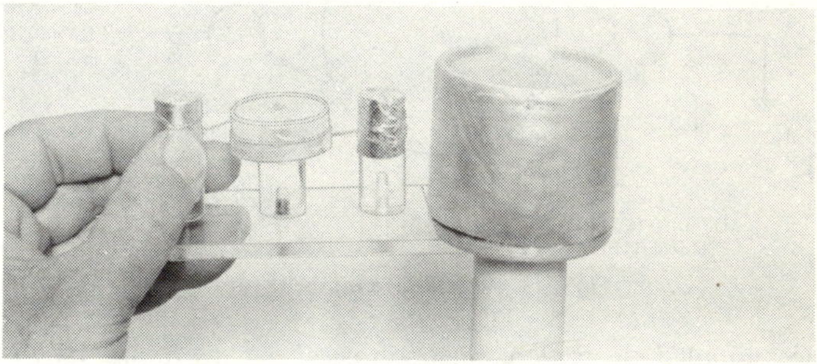

Fig. 5-8. Testing the plastic-disc motor with a Van de Graaf generator.

generator, and the other foil electrode is placed near a spark-gap assembly (see Chapter 2) metal ball that is grounded. The speed of the motor can then be adjusted by varying the spacing of the metal ball from the foil electrode (close spacing causes fast rotation of the motor).

The motor will also operate by connecting the foil electrodes to a dual polarity generator (such as the Varley). This is done by spacing the spark-gap metal balls far enough apart so that they are inoperative and the full energy of the generator is applied to the motor.

Plastic-Disc Motor and Plastic-Drum Motor Theory

Look at the drawing of Fig. 5-9. One foil electrode is connected to ground, and the other foil electrode is connected to (or in an electrostatic field of) a high-voltage electrostatic source, having either positive or negative polarity.

The corona discharge from the point of the needle on each foil electrode sprays electrical charges to a surface area of the plastic disc (wire brushes spray charges on the surface of the plastic-drum motor), and the plastic becomes charged with the same polarity charges as the adjacent foil electrode. Thus, the plastic surface area is repelled away from the needle and rotates the disc.

As the plastic surface area approaches the needle on the oppositely charged foil electrode, the plastic becomes neutralized because the electrical charges discharge to ground. Additional surface areas become charged and repelled from the high-voltage foil electrode, repeating the process and continuing to rotate the disc.

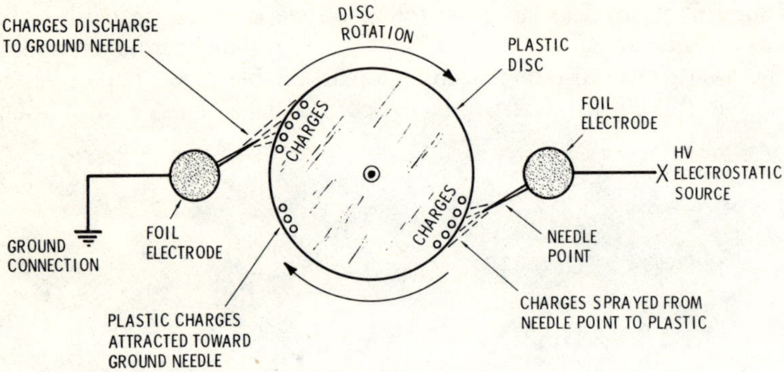

Fig. 5-9. Plastic-disc motor theory.

The grounded electrode connection can be made to an electrostatic source of opposite polarity to that of the one connected to the other foil electrode. The operation is then essentially the same, except that both needle points spray electrical charges after the plastic surface is neutralized.

PLASTIC-DRUM MOTOR

Construction of the Plastic-Drum Motor

As shown in Fig. 5-10, this electrostatic motor is built on 3¾- by 1½- by 3/16-inch plastic baseboard H (Fig. 5-11). The operation of this motor is dependent upon proper balancing of drum C and correct fit of bearings, B. These B bearings are 6-32 nylon machine screws, obtainable from electronics mail-order houses. Sections of threaded nylon or soft plastic rod can also be used.

Begin construction by selecting a plastic container with straight sides, suitable for the drum. We cut the bottom section of a 2-ounce pill container into the 1⅝-inch-diameter by 1-inch-long drum for our model, the exact size of the drum is not critical.

Shape the ends of a 1½-inch-long by ¼-inch-diameter plastic rod into pivot points to make motor shaft G. (A lathe works best for this shaping, but you may also grind the ends while the rod is chucked in a drill press.) Carefully drill a tight-fitting hole in the exact center of the drum and insert the motor shaft. Set aside the assembly.

Cut the two ¼-inch-square by 1¾-inch-high motor shaft support A sections, and drill and tap a centered 6-32 hole, ¼ inch from one end of each section. Drill a small centered hole in the end of each nylon machine screw, B, to fit the point of the motor shaft and insert a nylon machine screw into the threaded hole of each end section.

Locate and cement the end sections in place on the baseboard to fit the motor shaft as shown in the drawing. Adjust the nylon screws so that the motor shaft turns freely and center the drum along the motor shaft.

Fabricate two brush support sections, F, by drilling and tapping one end on each of two ½-inch-diameter by 1½-inch-long plastic rods. Install a solder lug with ½ inch long stranded wires connected to its end, and a 1 inch long machine screw and nut in the tapped hole of each brush support section. Tighten the nut so that the solder lug is flush with the surface of the plastic.

Bend one solder lug downward and the other one upward. Then cement the brush support sections in place on the baseboard so that the brush wires are close to the drum—do not allow the brushes to touch the drum surface (Fig. 5-11).

Make sure that the drum and motor shaft assembly is balanced and turns freely. Then cement the drum in place on the shaft.

NOTE: The operation of the motor is the same as the plastic-disc and two-brush motor. It will operate well with the Van de Graaff generator and a ground connection, as shown in Fig. 5-12.

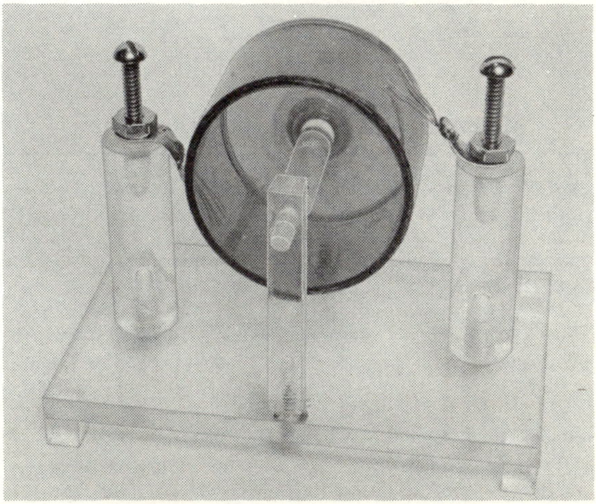

Fig. 5-10. Plastic-drum motor project.

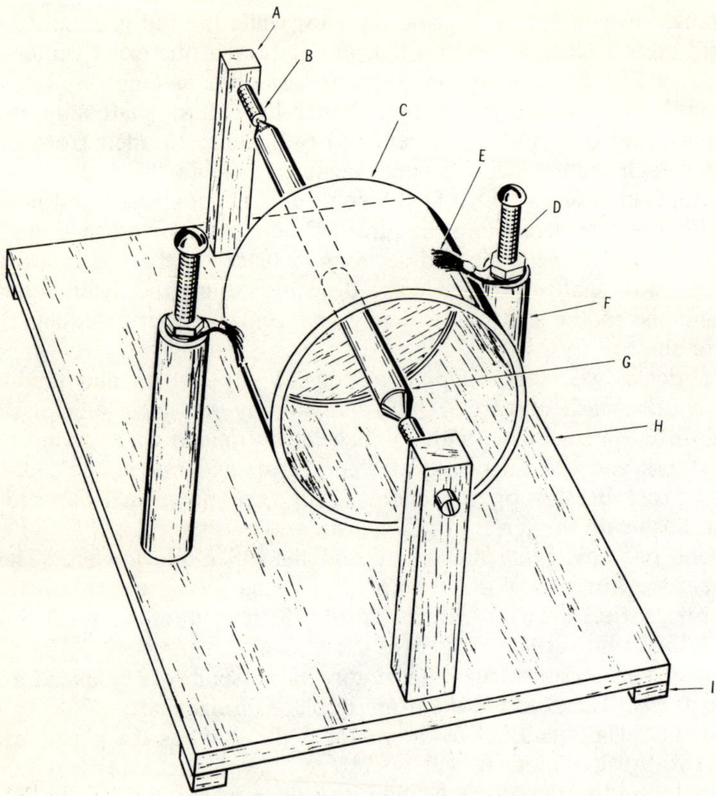

Fig. 5-11. Plastic-drum motor plans.

Materials List for Plastic-Drum Motor

A—Plastic section ¼″ square × 1¾″ high (with tapped hole for B)	E—Solder lug with ½″ long stranded wires connected to end
B—6-32 nylon screw with end drilled for point of G (See text.)	F—Plastic section ½″ diameter × 1½″ high
C—Plastic drum (section of plastic bottle) 1⅝″ diameter × 1″ long (See text.)	G—¼″ diameter plastic rod approximately 1½″ long and pointed at ends (See text.)
D—6-32 machine screw 1″ long (with nut)	H—Plastic base approximately 3¾″ high × 1½″ × 3/16″
	I —¼″ square plastic feet

Fig. 5-12. Testing the plastic-drum motor with one electrode close to a Van de Graaff generator and the other electrode connected to ground.

TWO-BRUSH DISC MOTOR

Construction of the Two-Brush Disc Motor

This electrostatic motor is built on a 3/16-inch-thick plastic baseboard approximately 4 inches by 2½ inches (Fig. 5-13). The assembly drawing of Fig. 5-14 shows the components involved in the construction of the motor. As in the plastic-drum motor project, the proper operation of this motor is dependent upon the balance of disc C and the correct fit between motor shaft H and bearing screws B. The bearing screws are nylon. (See the plastic-drum project.)

Begin construction by cutting plastic disc C from a section of 3/16-inch-thick sheet plastic. (Either a flycutter, chucked in a drill press, or a lathe can be used.) Drill a ¼-inch center hole to fit motor shaft H. Cut and cement aluminum foil sections on one side of the plastic disc, as shown in Fig. 5-14. Make sure that the foil sections are evenly spaced around the disc.

Shape the ends of a ¼-inch-diameter by 1-inch-long plastic rod into pivot points (as detailed in the plastic drum project to make motor shaft H. Fit the motor shaft into the plastic disc).

Cut two ¼-inch-square by 1¾-inch-long motor shaft support end sections labeled A. Drill and mount nylon bearing screws 1 inch from the top ends, as detailed in the plastic-drum motor project.

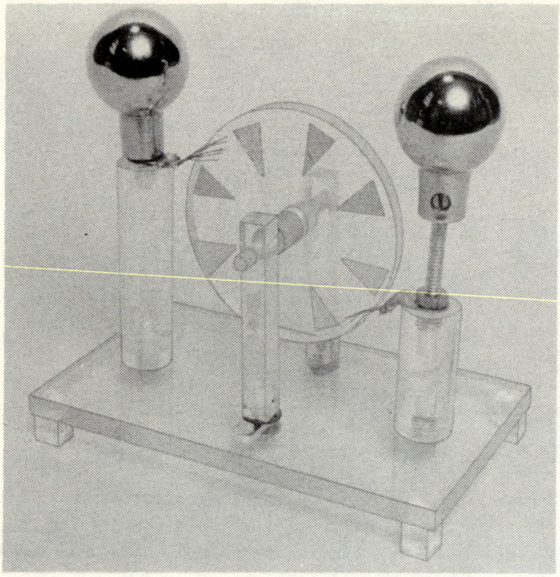

Fig. 5-13. Two-brush disc motor project.

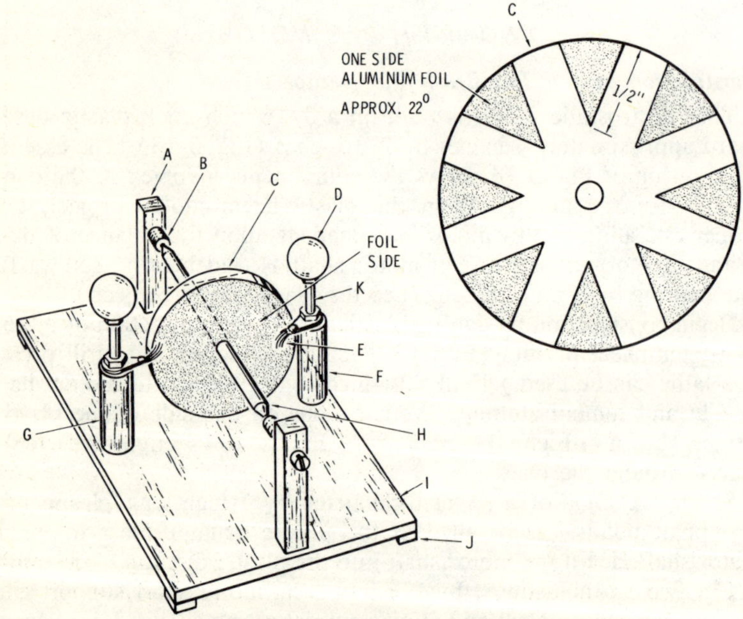

Fig. 5-14. Two-brush disc motor plans.

Locate and cement the end sections in place on the baseboard to fit the motor shaft, as shown in the drawing, or drill and tap the base of the end sections and mount the sections with machine screws in slotted holes on the baseboard, as in our model. Adjust the nylon screws so that the motor shaft turns freely and center the plastic disc along the shaft.

Fabricate brush support section G by drilling and tapping one end of a ½-inch-diameter by 1¾-inch-long plastic rod. Install a solder lug with ½-inch-long stranded wires connected to its end, and a machine screw, nut, and metal knob D. Tighten the nut and metal knob down, so that the solder lug is flush with the plastic rod surface. Cement brush assembly G to the baseboard near the foil side of the plastic disc and bend the solder lug upward, so that the brush wires are close to, and pointing toward, the top edge of the disc (as shown in the drawing).

Materials List for Two-Brush Disc Motor

A—Plastic section ¼" square × 1¾" high, with tapped hole for B	F—Plastic section ½" diameter × 1" high
B—6-32 nylon screw, with end drilled for point of H (See text.)	G—Plastic section ½" diameter × 1¾" high
C—Plastic disc 2" diameter × 3/16" thick, with foil segments (See drawing.)	H—¼" diameter plastic rod approximately 1" long and pointed at ends (See text.)
D—Metal knob, and machine screw and nuts (See text.)	I—Plastic base approximately 4" × 2½" × 3/16"
E—Solder lug with ½" long stranded wires connected to end (See text.)	J—¼" square plastic feet

Plastic rod F is shorter than plastic rod G, but the threaded rod is longer. Fabricate brush assembly F the same as G, except with a ½-inch-diameter by 1-inch-long plastic rod. Adjust the height of the metal knob so that it will be the same height as the metal knob on G when the brush support section is installed. Cement the brush section to the baseboard, as shown in the drawing, so that the brush wires are close to, and are pointing down toward, the bottom edge of the foil side of the disc.

Make sure that the plastic disc and motor shaft turn freely and that they are balanced. Also, check to see that the brush wires are close to, but not touching, the foil surface of the disc; then, cement the disc to the motor shaft.

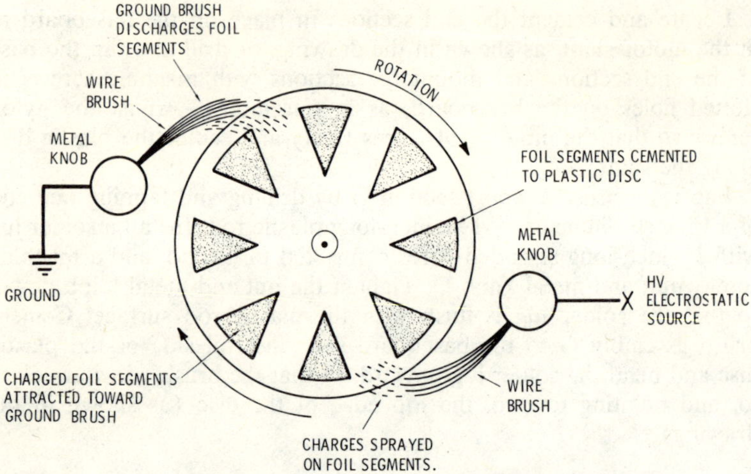

Fig. 5-15. Two-brush disc motor theory.

Two-Brush Disc Motor Theory and Operation

As shown in Fig. 5-15, one metal knob is connected to ground, and the other metal knob terminal of the motor is connected to a high-voltage electrostatic source. The theory of this motor is the same as that of the plastic-disc motor and plastic drum motor, except

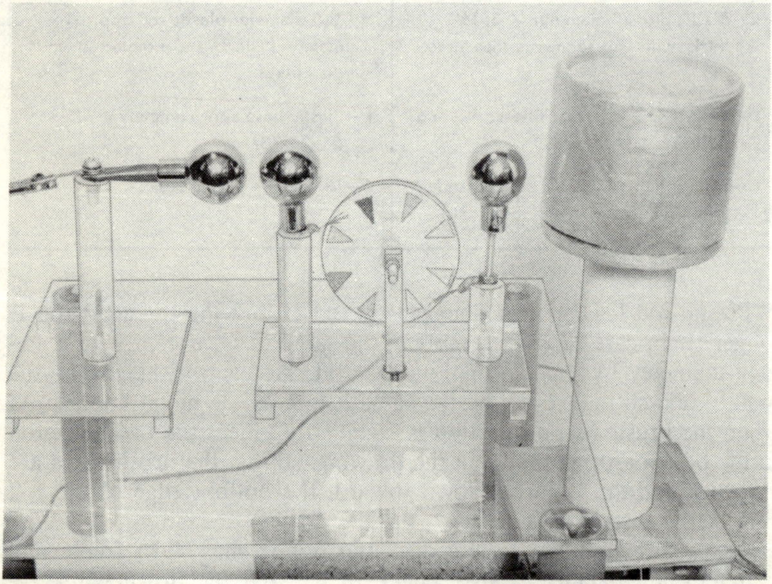

Fig. 5-16. Testing the two-brush disc motor with a Van de Graaff generator and a grounded spark-gap assembly.

that the sprayed charges from wire brushes are carried by the aluminum foil segments in the two-brush disc motor.

The operation of this motor is also the same as the plastic-drum motor, and it can operate between two high-voltage electrostatic sources of opposite polarity, instead of one high-voltage source and ground (Fig. 5-16). It can also be connected between the electrodes of a high-voltage electrostatic source that will deliver both polarities at the same time (such as the Varley generator).

FOUR-BRUSH DISC MOTOR

Construction of the Four-Brush Disc Motor

This electrostatic motor is more complicated than the other motor projects, and it requires more patience and care in its construction (Fig. 5-17). The motor shaft pivot points and nylon bearing screws

Fig. 5-17. Four-brush disc motor project.

and holes, B, are fabricated in the same way as in the plastic-drum motor project and the two-brush motor project. See these projects for detailed descriptions.

The motor is built on 2½- by 4- by 3/16-inch plastic baseboard G in Fig. 5-18, and 2¼- by 3¼- by 3/16-inch plastic backboard A, as shown in the drawing. Begin construction by fabricating plastic disc C and motor shaft J, as detailed in the two-brush motor project. The disc foil segments are shaped as shown in Fig. 5-18 and are cemented on both sides. The motor shaft is 1½ inches long.

Drill and tap a centered hole for the nylon bearing screw at a point 1¾ inches up from the bottom of the A section, as shown in the drawing. Temporarily place disc C over the nylon screw and locate the four threaded rods F, as indicated in the drawing. Cement section A to the rear of the baseboard as shown. (A stronger assembly can be made with two tapped holes and machine screws bolted to the baseboard in the bottom side of section A, as in our model.)

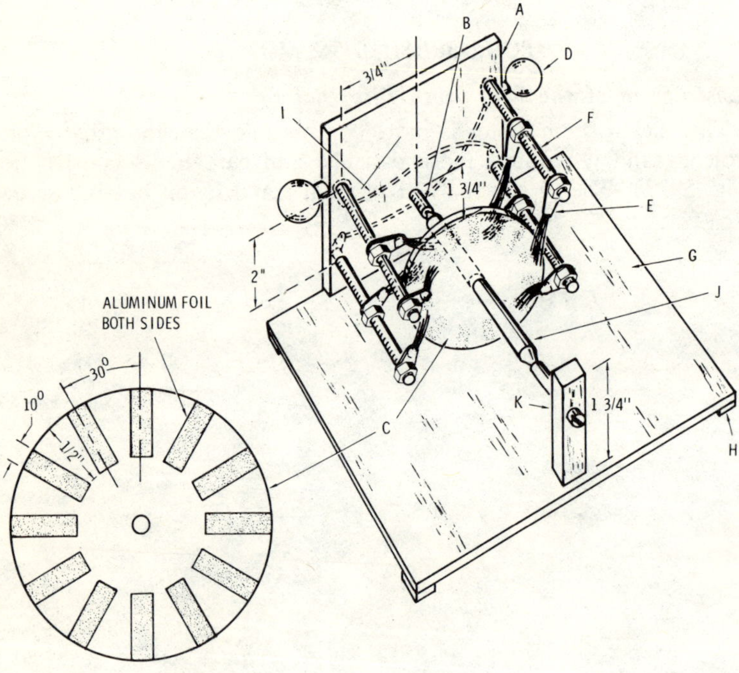

Fig. 5-18. Four-brush disc motor plans.

Fabricate motor-support plastic section K, as shown in the drawing, with a tapped hole for the nylon screw 1¾-inch up from the base. Mount section K with a tapped hole and machine screw or cement it to the baseboard. Install the disc and motor shaft assembly with the two nylon screws. Install the four threaded rods, F, with two solder lugs and ½-inch wire brushes E spaced with nuts on each side of the motor disc. Point the brushes in the directions shown.

Adjust the brushes so that they are close to, but not touching, the disc surface while the disc revolves. Mount two metal knobs, D, on the threaded rods with solder lugs mounted in threaded holes close to the top set of brush support rods F (one knob on each side, as shown in the drawing). Connect the knob solder lugs to the solder lugs of the nearby F rods. Then, connect the solder lugs of the brushes together with insulated wire as shown in the drawing.

Materials List for Four-Brush Motor

A—Plastic section 2¼″ × 3¼″ × 3/16″	F—1½″ long 6-32 threaded rod and nuts
B—6-32 nylon screw with end drilled for point of J (See text.)	G—Plastic base 2½″ × 4″ × 3/16″
C—Plastic disc, 2″ diameter × 3/16″, with foil segments (See drawing.)	H—¼″ square plastic feet
D—Metal knob, machine screw and nuts (See text.)	I—Hookup wire (See text.)
E—Solder lug with ½″ long stranded wires connected to end (See text.)	J—¼″ diameter plastic rod approximately 1½″ long and pointed at ends (See text.)
	K—Plastic section ¼″ square × 2″ high

Four-Brush Motor Theory and Operation

This motor uses four dual brushes connected as shown in Fig. 5-19. The dual brushes are connected together electrically by their mounting rods and are positioned so that each section of a dual

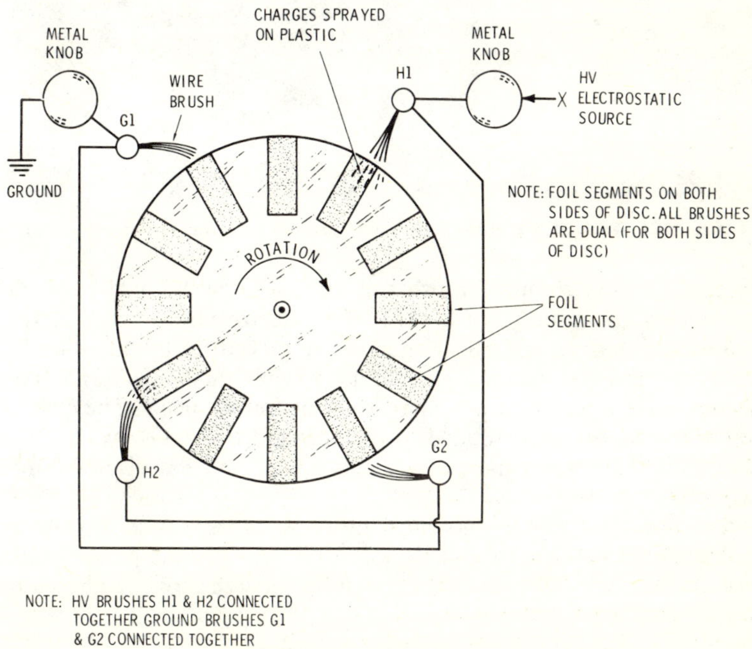

Fig. 5-19. Four-brush motor theory.

brush can spray charges on its own side of the rotating disc. The plastic disc has aluminum foil charge-carrying segments on both sides (Fig. 5-18).

As shown in Fig. 5-19, one metal knob terminal is connected to ground, and the other metal knob is connected to a high-voltage source of electrostatic energy, such as the Van de Graaff generator. The dual brushes are connected together as shown in the drawing. (For simplification, only one side of the disc and its dual brushes are shown.)

As dual brushes H1 and H2, connected to the high-voltage source, spray charges on the foil segments on both sides of the disc, the disc

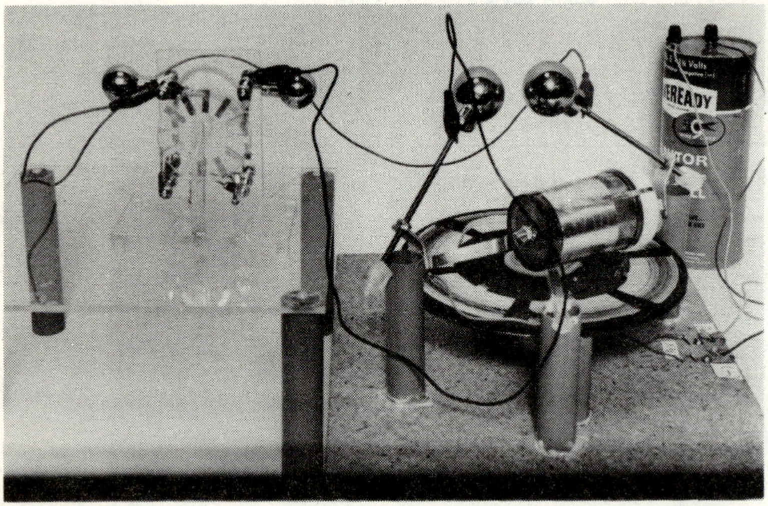

Fig. 5-20. Testing the four-brush motor project with the Varley machine.

is repelled away from the brush and attracted toward dual brush G2 (connected to ground brush G1). The electrostatic charge is then neutralized and, as additional charges are picked up by the other foil segments, the disc continues to rotate toward dual brush H2 (connected to H1), which sprays charges on the foil segments. The charged foil segments are then rotated toward ground brush G1, where their charge is neutralized; the cycle begins again, and the disc continues to rotate.

The operation of this motor is the same as that of the two-brush motor, and it will work the same way with two opposite polarity high-voltage sources (Fig. 5-20) as well as between one high-voltage source and ground.

CHAPTER 6

Electrostatics and Radiation

Perform the following experiment: charge an electroscope with either a positive or negative charge, and then hold a lighted match near the metal knob of the electroscope. You will observe that the leaves of the electroscope discharge (move together). This effect is due to the ionization of the air close to the metal knob produced by the burning gasses of the lighted match. The ionization makes the air conductive, and the air neutralizes or rapidly leaks off the charge of the electroscope.

EARLY FORMS OF RADIATION DETECTION

Ionization is also produced by the radiation from radioactive minerals. In 1902, Madame Curie and her husband used electroscopes to measure the intensity of radioactive minerals in their successful search for radium. Until the introduction of modern electronic detection devices, various forms of electroscopes were used by scientists in their investigation of radioactivity. The intensity of radiation was determined by the rate of discharge of an electroscope.

In 1908, German physicist Hans Geiger devised an early form of radiation detector that utilized the effects of ionization of gas to operate an electroscope (Fig. 6-1). In the detector shown in Fig. 6-1, the high-voltage battery provides a voltage that is just below the ionization point of the gas in the ionization chamber. When radiation (cosmic rays, etc.) enters through the metal foil "window" a few ions are produced due to the effect of the radiation on some of the gas atoms. These ions cause a momentary current flow in the chamber and through the resistance. The voltage change across the resistance causes the electroscope leaves to momentarily change position. This change in position is observed through a magnifying device inserted into the shielded electroscope.

In the process of investigation of radioactivity, very sensitive electroscopes were developed, and scientists noticed that these electroscopes would slowly discharge over a period of time by what was called "natural leakage." At first it was thought that natural radiation from the earth caused this leakage. But, after shielding experiments with lead and charged electroscopes carried up to the stratosphere in balloons, it was found the leakage still occurred. In 1911, Austrian physicist, Victor F. Hess discovered that radiation from outer space was discharging the electroscopes. In 1925, American physicist Robert A. Millikan performed further investigation by sending a series of balloons up to the stratosphere with electroscopes and cameras. He

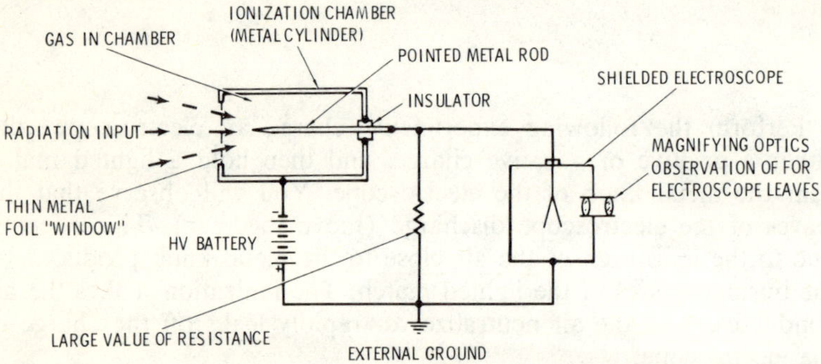

Fig. 6-1. Early form of radiation detector.

found that radiation penetrated and discharged his instruments. Millikan named these radiations "cosmic rays." Scientists have continued to investigate cosmic rays with various types of instruments.

Investigation of these cosmic rays has shown that they are high-energy particles which strike air molecules in the stratosphere and cause secondary high-energy particles from the air molecules to send "showers" of radiated particles to the earth below.

In 1912, Scottish physicist C. T. R. Wilson invented the "cloud chamber." This device produces a visible track of minute droplets along the path of a single high-energy ionizing particle. Thus, the charged particles emitted by radioactive sources and cosmic rays could be made visible in the form of tracks through a container of supersaturated vapor.

The Wilson cloud chamber used a mechanical expansion to create the conditions necessary for making the particle tracks visible. The cloud chamber described in this chapter is a simplified type based upon a design by scientists of the Brookhaven Laboratories (a national atomic research center).

ELECTROSCOPE TEST FOR RADIOACTIVITY

The electroscope was used by early experimenters to test for radioactivity, as shown in Fig. 6-2. This was done by first charging a shielded electroscope and then determining the time that it took for the electroscope to normally discharge itself to a given angle of the foil leaves (Fig. 6-2A).

> NOTE: An electroscope will normally discharge itself over a period of time because of insulation leakage, humidity of the air, and normal background radiation.

A radioactive specimen was then placed into the shielded electroscope (Fig. 6-2B), and the electroscope was recharged. The electroscope was then timed for the foil leaves to discharge to the same given

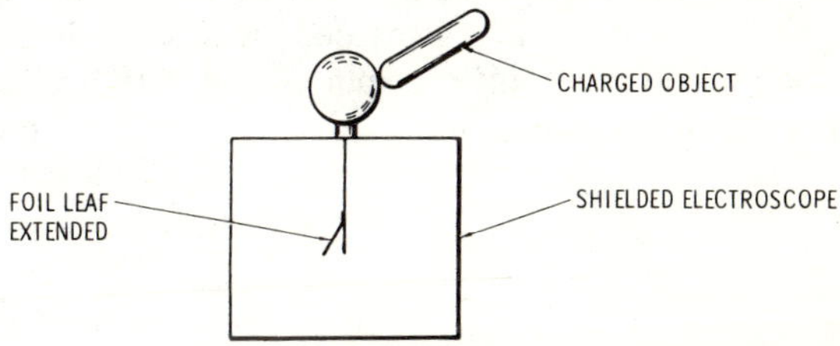

(A) Electroscope is charged and time measured for foil leaf to drop to discharged position.

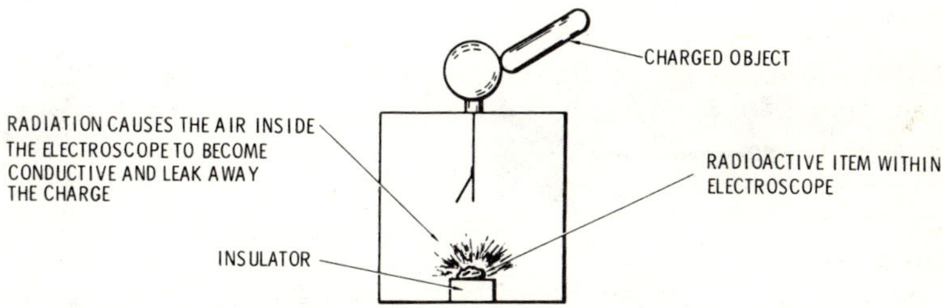

(B) Electroscope is recharged and time measured for foil leaf to drop to discharged position, in the presence of radiation. The two measured times are compared to determine the degree of radioactivity.

Fig. 6-2. Electroscope test for radioactivity.

angle (for example, 10 degrees). The electroscope would discharge in a shorter time (dependent upon the amount of radioactivity in the specimen) than it did for the original time period.

Some molecules of air are struck by the fast-moving particles from the radioactive specimen. Electrons are either lost or gained by the air molecules and become either positive ions (minus electrons) or negative ions (additional electrons). These ions are attracted by the charged foil leaves of the electroscope. If the electrostatic charge of the electroscope is negative, the foil leaves will attract the positive ions; if the charge is positive, the leaves will attract negative ions. The ions (of either polarity) will neutralize the electrostatic charge on the foil leaves over a period of time. With higher radioactivity, more ions will be emitted, and the electrostatic charge will be neutralized faster.

CLOUD CHAMBER

Construction of the Cloud Chamber

This project requires a weak, radioactive alpha source, such as a speck of the radioactive paint from a dial or hand on a watch or a clock that glow in the dark. If a discarded clock is available, you can scratch the paint with the point of a pin coated with a bit of cement. NOTE: Only the paint that glows in the dark *without* previous exposure to light is suitable. Paint that requires previous light exposure is phosphorescent paint, not radioactive paint, and cannot be used for this project. If no radioactive paint is available, an alpha source can be obtained at a science hobby store or from a science laboratory mail-order house. Observe suitable *precautions* even when working with this type of weak, radioactive material. Do not *touch* the radioactive paint with your fingers. Do not *inhale* air from any container that holds the source. Wash your hands thoroughly after any contact. Keep the source in a sealed plastic container when not in use.

Fig. 6-3. Cloud chamber project.

As shown in Fig. 6-3, a clear plastic food storage jar approximately 3¾ inches in diameter by 3 inches high is used for the cloud chamber. The jar is used upside down, with the plastic lid as the base. There-

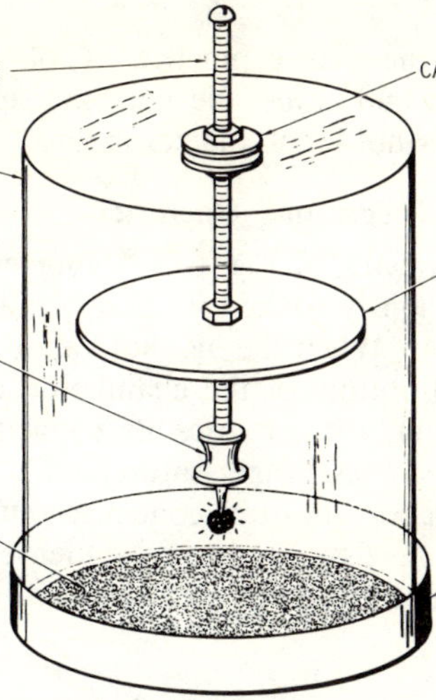

Fig. 6-4. Cloud chamber plans.

fore, select a jar with a flat lid that has no projections and can be placed in a level position.

Begin construction by making a center hole in the jar bottom for the machine screw (Fig. 6-4). (You can use a heated ice-pick to make the hole.) Cut two 1 inch diameter cardboard washers. Cement one washer on the inside of the jar over the hole, and the other one over the hole on the outside. Cut a 2¾-inch cardboard disc and mount it in the jar with the 3-inch machine screw and nuts. Use nuts on both sides of the 1-inch washers.

Cut a disc of black paper to fit the inside of the jar lid and cement the paper in place, as shown in the drawing. Mabe sure that the paper surface is smooth, with no bumps or wrinkles.

Drill and tap a hole in the top of a plastic head push-pin to fit the end of the machine screw. (This type of pin can be obtained at an office supply store.) Place a small amount of plastic cement on the tip of the pin, and then scrape the radioactive paint of the numerals or hands of a discarded clock with the pin point (while holding the plastic end only). Mount the pin on the end of the machine screw

Materials List for Cloud Chamber

Machine screw and nuts	Radioactive source (See text.)
Cardboard	Black paper
Clear plastic jar and lid	Dry ice
Plastic top pushpin	Light source (See text.)

by holding only the plastic pin head (do not touch the pin point). Install the jar lid and adjust the machine screw as necessary so that the pin point does not touch the cardboard.

Cloud Chamber Theory and Operation

Look at the drawing of Fig. 6-5. Vapor evaporates from the cardboard disc (previously soaked in alcohol) and mixes with the air in the cloud chamber. When the alcohol vapor-air mixture settles downward to the cold bottom of the chamber (chilled by the dry ice slab under the bottom), the air loses its capacity to hold much alcohol vapor; the air becomes "super-saturated."

The alcohol vapor does not condense out of the cold air immediately, because it needs particles to condense on (like water vapor in the sky needs particles to condense on to form rain drops).

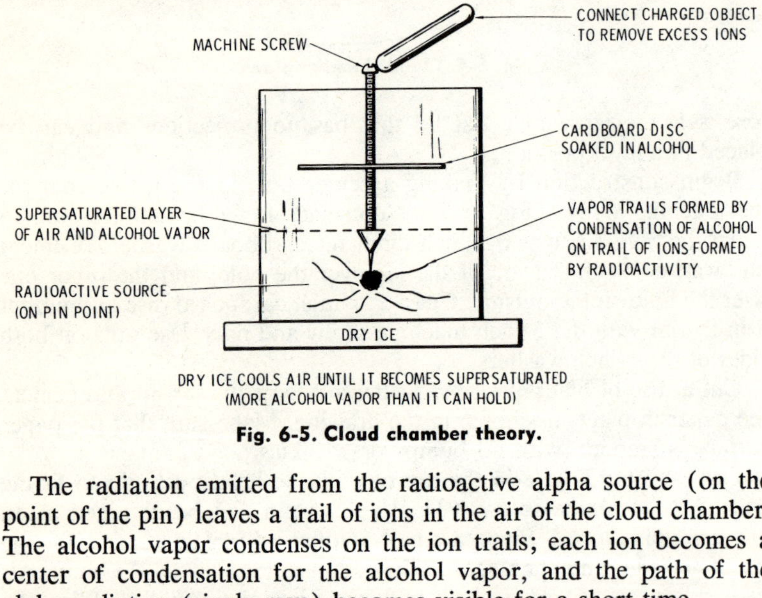

Fig. 6-5. Cloud chamber theory.

The radiation emitted from the radioactive alpha source (on the point of the pin) leaves a trail of ions in the air of the cloud chamber. The alcohol vapor condenses on the ion trails; each ion becomes a center of condensation for the alcohol vapor, and the path of the alpha radiation (single rays) becomes visible for a short time.

The vapor trails will be difficult to see after a period of the cloud chamber operation because of the large generation of excess free ions. These ions will eventually interfere with the production of vapor trails. The excess ions can be removed by first connecting a negatively charged object to the head of the machine screw (top of the cloud chamber), which will attract the positive ions to the machine screw. Then, connect a positively charged object to the machine screw to attract the negative ions.

The vapor trails can best be observed in a dark room with a narrow beam of light focused under the pin point and across the top of the black cardboard on the cloud chamber bottom. This beam of light can be produced by a powerful flashlight (Fig. 6-6) or slide projector. The vapor trails of the alpha radioactive source are usually straight lines radiating outward from the pin point.

You may be able to observe secondary cosmic rays by the effects of these rays in collision with a thick layer of metal. To observe these effects, remove the radioactive alpha source and mount a heavy metal washer (iron, steel, etc.) in the same place on the end of the machine screw. Patience and time are required to observe the "showers" of vapor trails from the bottom of the metal disc when the metal disc is hit by a cosmic ray.

Ethyl alcohol (or the denatured type) and some types of rubbing alcohol will be satisfactory for soaking the cardboard disc. If soft cardboard that will hold the alcohol is not available, try heavy blotting paper.

Fig. 6-6. Operation of the cloud chamber project using a flashlight for a light source.

Dry ice may be difficult to get in small quantities. Try a local ice cream dealer or camping supplies dealer. You can also check with a dry-ice manufacturer and have the dry ice cut up into small slabs. These dry ice sections can be kept for a few days in a portable insulated ice chest. (Ask the dry-ice manufacturer for details on dry ice storage.) Do not handle the dry ice with bare hands; use heavy gloves. When operating the cloud chamber, place a wooden box between the dry ice and the table-top or hold the dry ice in a wooden bowl. The dry ice should be cut (use a hacksaw) so that the top of the dry ice slab is in close contact with the bottom of the cloud chamber.

TYPES OF VAPOR TRAILS

Alpha Particle

This is a heavy particle equivalent to the nucleus of a helium atom (two protons and two neutrons), and it has a positive electrical charge. As the particle moves through the cloud chamber, it attracts electrons from air molecules. These air molecules become ions and the alcohol vapor condenses on these ions, forming straight and thick visible vapor trails radiating outward from the radioactive source.

Beta Particle

This particle is an electron emitted from a radioactive source. The beta particle has a negative electrical charge and, being very lightweight, they travel at almost the speed of light. The beta vapor trails are thin and wavy because the lightweight particles are bounced about and are much harder to detect than the alpha vapor trails.

CHAPTER 7

Modern Electrostatic Devices

Until the time of Benjamin Franklin, the average man either did not think at all about the mysterious forces of electricity, or he thought it was a form of magic. In particular, the mighty force of lightning was regarded with superstitious awe and fear. Franklin's invention of the lightning rod, in 1747, was the first, practical, electrical device for the home that the average man could comprehend. Benjamin Franklin showed the world how to protect property from lightning.

After the first electrostatic device for the home was invented, the average man had little to do with electrostatic devices until modern times. An electrostatic pollen-removal device for those with an allergy can be found in homes today. The lightning rod is still on many homes, practically unchanged in design. The Leyden jar-type capacitor is in all radios and television sets.

Industry has utilized electrostatic devices for smoke removal, paint spraying, manufacturing of sandpaper, and in the Xerographic copy and reproduction process.

POSSIBLE FUTURE USES OF ELECTROSTATICS

The Electret

In 1925, Japanese physicist Dr. Eguchi discovered that a mixture of waxes (such as 45% carnauba wax, 10% white beeswax, and 45% white resin), heated and then allowed to cool in a strong electrostatic field, will become electrically polarized. The polarization maintains a positive charge on one surface and a negative charge on the other surface. Thus, a polarized wax section is the electrostatic equivalent of a permanent magnet and is called an "electret." The electrical charges of the electret may last for a considerable period of time. It is believed that the wax acts as a dielectric in a capacitor, and its molecules become aligned when the wax is hot. Then, when the wax is cold, the electrical charges are "frozen" in place.

Recently, experiments have been made in forming electrets from plastic instead of wax, and future uses for the plastic electret may include microphones and sound-powered telephones. The electret may also be used in electrostatic speakers. See "The Not Altogether Forgotten Electret" by Herbert and Phyllis Cohen in the March 1969 issue of *Popular Electronics* for more information and details on constructing an electret.

Control of Weather by Electrostatics?

Around the beginning of the century, inventor Nikola Tesla tried to start rain storms by shooting gigantic electrical sparks into the sky from his Tesla coils. Modern-day experiments are cloud seeding with silver iodide smoke. The future may involve the release of ions to start rain in clouds by corona discharge from electrostatic generators carried in aircraft. Charging of the atmosphere with ions may have side effects, because negative and positive ions seem to cause emotional feelings in some people.

Possible Electrical Power from the Earth's Electrostatic Field

The earth has an electrostatic field surrounding it. This field is between the surface of the earth and the ionosphere (a layer of ionized gas in the stratosphere). The voltage of the electrostatic field (measured in respect to the earth) increases many thousands of volts, until it reaches a potential of hundreds of thousands of volts at the ionosphere.

This electrostatic energy may be used in the future by means of ionizing rays from the ground acting as high antennas to collect the energy. Electrostatic motors (as in Chapter 5), with improved high-voltage insulation and operation, may be used in place of the conventional electric motors.

EXPERIMENT TABLE

The experiment table is shown in Fig. 7-1. The dimensions of the table top are not critical (our model is 8 inches square by ⅛-inch-thick), but the height of the table legs should be the same as in Fig. 7-2 (4 inches high by 1 inch in diameter) to properly coordinate with the experiment projects.

The tabletop and legs should be constructed of plastic with good electrical qualities (we used clear sheet plastic for the top and PVC plastic tubes for the legs).

Construct the table as shown in Fig. 7-2. If plastic tubing is used for the table legs, cement short sections of plastic rod inside the tubing

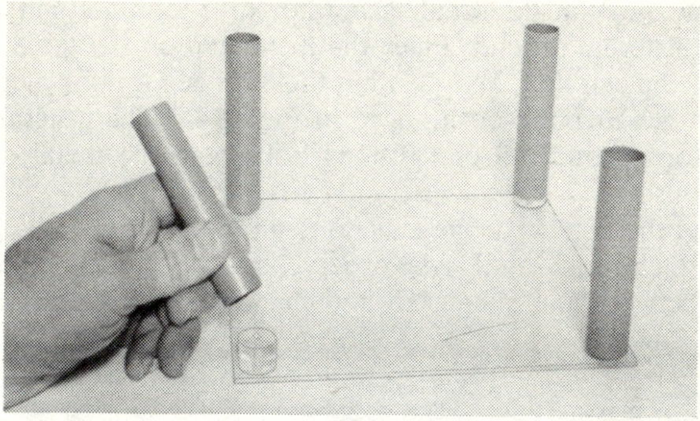

Fig. 7-1. Assembling the plastic experiment table.

and then cement or bolt (with plastic bolts) the legs to the table corner. Make sure that the legs are the same length; the table should be level.

SIMULATED LIGHTNING ROD EXPERIMENT

During Franklin's investigation of electricity, he noticed that a sharply pointed, grounded, metal rod would draw off the electrical energy from a charged object without a spark discharge. This observation was the foundation of his invention of the lightning rod.

Since an actual experiment with a lightning rod and an electrical storm would be dangerous, a simulated lightning rod experiment is given in Fig. 7-3. It will show how a lightning rod operates.

You will need a small Van de Graaff generator, a plastic experiment table (Fig. 7-1), and a 1¾-inch-high by approximately 2½-inch-diameter metal can bottom with a rounded edge. (We used a beer can similar to the one used in the Van de Graaff generator project of Chapter 2.)

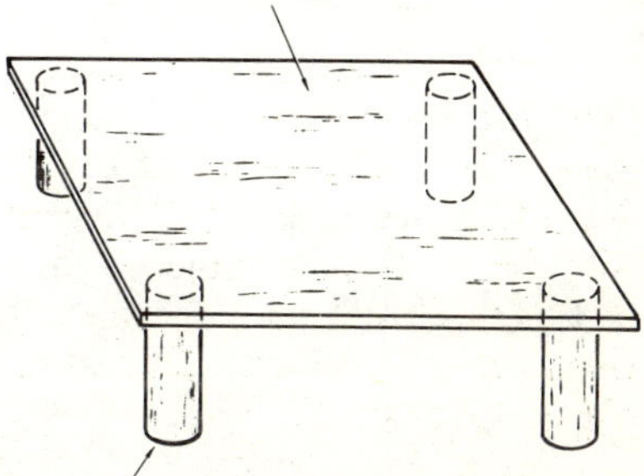

Fig. 7-2. Plastic experiment table plans.

Connect the Van de Graaff generator GND terminal and the metal can to an external ground. Place the metal can on the table (with the bottom of the can up), and position the Van de Graaff generator over the can in a horizontal position, as in Fig. 7-4. (The generator represents a storm cloud full of lightning bolts, and the metal can represents a house.)

Connect the battery to the generator motor and move the generator over the metal can until you observe sparks (simulated lightning) jumping between the metal can and the generator. Note the farthest

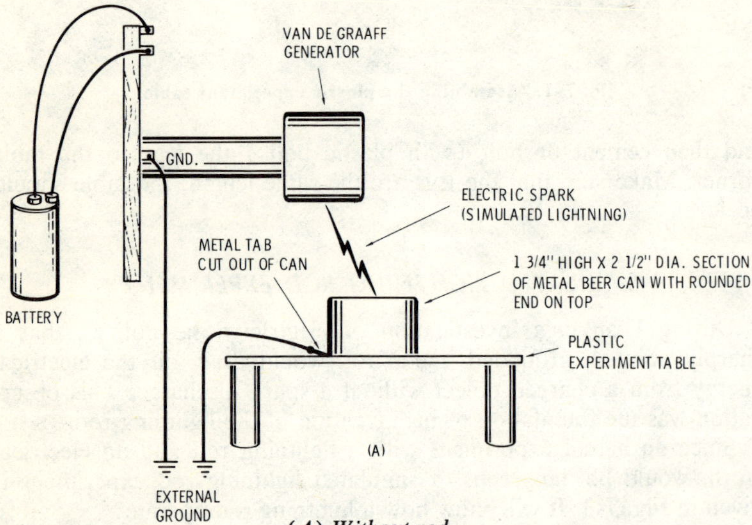

(*A*) *Without rod.*

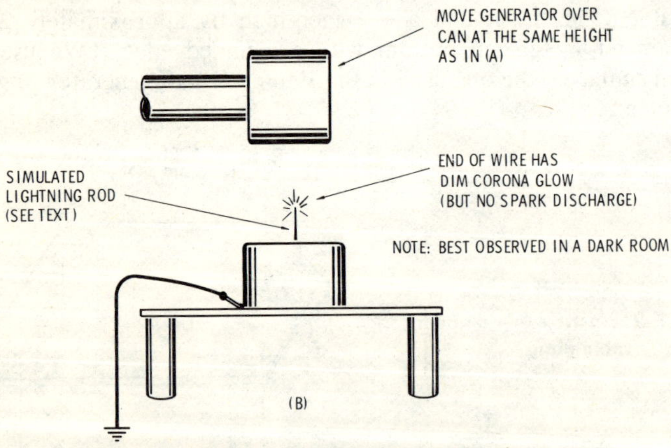

(*B*) *With simulated lightning rod.*

Fig. 7-3. Simulated lightning rod experiment.

distance that sparks will jump when the generator is directly over the can.

Tape a length of No. 22 bare, solid copper wire to the center of the top of the metal can (make sure that there is a good electrical connection) and bend up ¾ inch of the free end of the wire. Clip the end of the wire diagonally so that it is pointed (the wire represents a lightning rod).

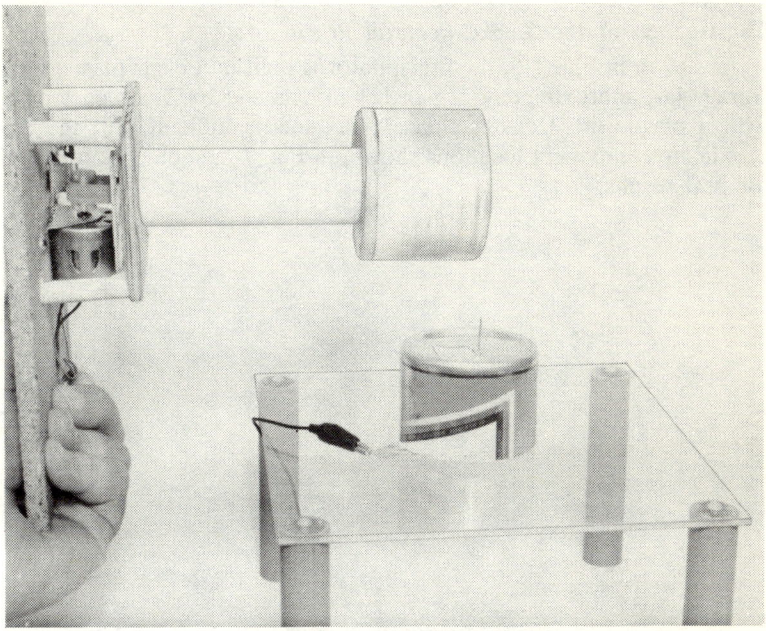

Fig. 7-4. Performing the simulated lightning rod experiment using a Van de Graaff generator.

Move the generator over the metal can at about the same height noted in the previous experiment, and observe that there is no spark discharge. The sharply pointed wire has drawn off the electrostatic energy from the generator by corona discharge (but without a spark).

Repeat this experiment in a dark room and you will notice that the

Materials List for Simulated Lightning Rod Experiment

No. 22 solid copper, bare wire	Power source Van de Graaff Generator (See text.)
Metal can section (See text.)	Connecting leads
Plastic experiment table	

end of the wire has a dim corona glow. Try the experiment with various lengths of wire (insulated, or not insulated, from the metal can). You can also try the experiment with model wooden houses and grounded miniature lightning rods of various shapes to determine the most efficient shape.

SMOKE REMOVAL PRECIPITATOR

Construction of the Smoke Removal Precipitator

As shown in Fig. 7-5, the precipitator is built in a clear plastic food storage jar, approximately 2½ inches in diameter by 2½ inches high, with a plastic lid. Construction should not be difficult, but the approximate component locations shown, in Fig. 7-6, should be followed for best results.

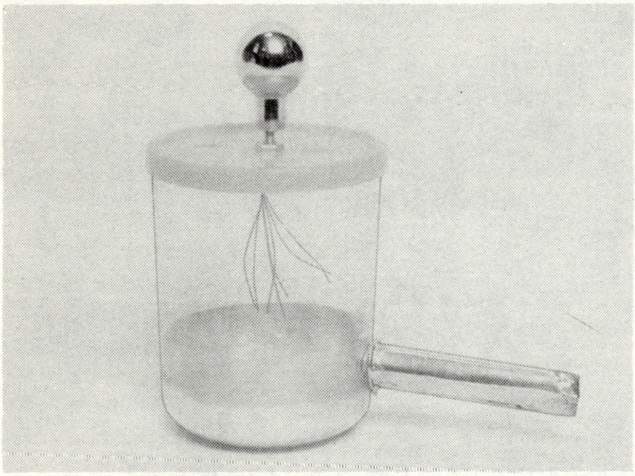

Fig. 7-5. The smoke removal precipitator unit.

Begin construction by cutting a ½-inch hole in the jar side near the bottom, as shown in Fig. 7-6, to fit the 2½ inch long by ½ inch diameter metal tube. The tube in our model was made from a section of tin-can sheet steel rolled into shape and soldered at the edges. Any type of lightweight metal that can be soldered, such as ½-inch tube copper, can be used instead.

Tape a section of household aluminum foil around the inside of the jar bottom and 1 inch up the sides. (Do not cover the tube hole.) Solder two 1-inch lengths of No. 22 solid copper wire to the edge of the metal tube. Insert the tube just inside the jar and bend the wires so that they make a good electrical contact with the aluminum foil. Wrap a few turns of wire around the tube just outside of the jar to keep the tube from moving. Solder the wire turns in place.

Solder a 2-inch length of stranded wire to a solder lug and fan the wire strands out so that each strand is separate and the ends are pointed outwards. Mount the solder lug with a machine screw and nut to a center hole in the jar lid (as shown in the drawing) and install a metal ball on the screw end. Cut two ¼-inch holes in the lid near the metal ball as shown. Place the lid on the jar.

Materials List for Smoke Removal Precipitator

Clear plastic jar and lid	Plastic tape
Metal knob	Tin can
Machine screw and nuts	Plastic table (See text.)
Stranded wire	Power source Van de Graaff Generator (See text.)
Aluminum foil	

Construction of the Smoke Generator

The smoke generator is made from a tin can approximately 2½ inches in diameter by 3¼ inches high, as shown in Fig. 7-7. The size of the tin can is not critical, and any size of can should work. Make sure that the tin can is clean and dry, with no inside coating.

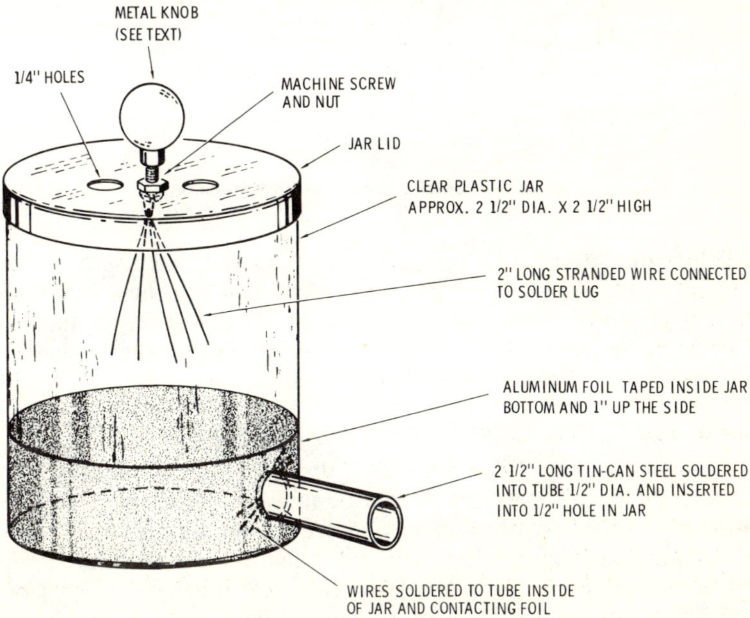

Fig. 7-6. Smoke removal precipitator unit plans.

Begin construction by cutting a ¾-inch hole in the side of the can near the bottom, as shown in the drawing. Then cut a ½-inch hole near the top and solder a short section of tin-can metal around the hole to fit around the metal tube of the smoke removal precipitator. If necessary, crimp the end of the precipitator tube so it will fit snugly inside the smoke generator. For best results, the two should join so that no smoke is lost from gaps in the joint. Do not solder them together.

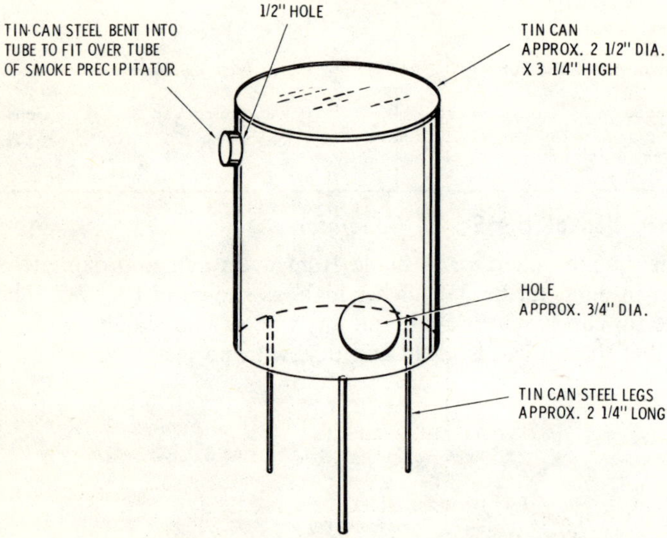

Fig. 7-7. Smoke generator unit plans.

Solder the lid (and the bottom) to the can and make sure that no air gaps exist. Cut sections of tin-can metal and roll them into three steel legs approximately 2¼ inches long. Solder them to the can bottom in the positions shown in the drawing. Adjust the leg lengths as necessary, so that the generator is level.

Smoke Removal Precipitator Theory and Operation

As shown in Fig. 7-8, smoke from the smoke generator unit flows through the metal tube, into the plastic jar, and out of the vent holes at the top of the jar. When an electrostatic generator (of either negative or positive output polarity) is connected to the metal knob at the top of the jar, the smoke particles become electrically charged (ionized) by contact with the electrified wire brush in the center of the jar. The charged smoke particles are then attracted to the grounded metal foil (some particles are attracted to the plastic sides of the jar), thus clearing the air of smoke.

The smoke generator unit (Fig. 7-7) should be placed close to the side of the experiment table (Fig. 7-2) and connected to the precipitator plastic jar, which is placed on the table (Fig. 7-9). A short section of paper is then set on fire and placed into the hole on the side of the smoke generator unit where it can smolder and give off smoke. (Try different types of paper for maximum smoke.) If necessary, the hole size can be altered for easier operation of the unit.

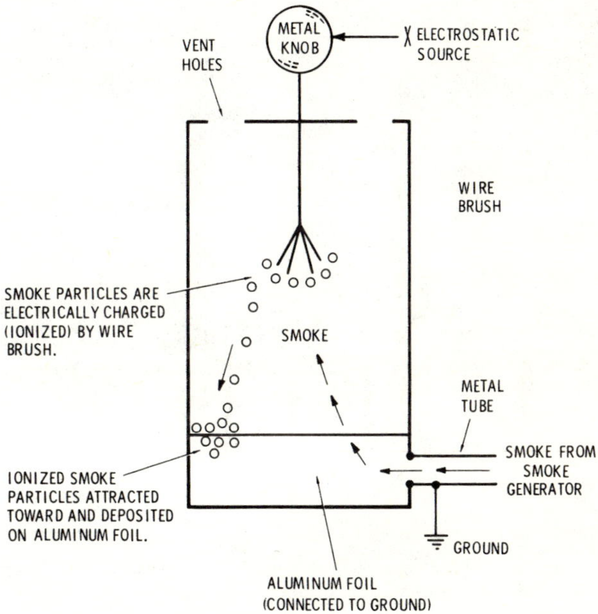

Fig. 7-8. Smoke removal precipitator project theory.

A Van de Graaff generator can be used for the electrostatic source, or you can experiment with the electrophorus and its charge carrier contacted to the precipitator metal knob for short term removal of smoke.

ELECTROSTATIC SPEAKER

Construction of the Electrostatic Speaker

Our model of an electrostatic speaker is built in a wooden picture frame designed for 8- by 10-inch pictures (Fig. 7-10). The size of the picture frame is not critical (larger frames will be more efficient at lower frequencies).

Cut three sections of household aluminum foil that will fit inside the frame. The foil should have a minimum of approximately ½ inch of space between the foil edge and the wooden frame. The foil sections should have tabs located as shown in Fig. 7-11. Cut four sections of plastic wrap to fit inside the picture frame with enough overlap to prevent the foil sections from touching each other.

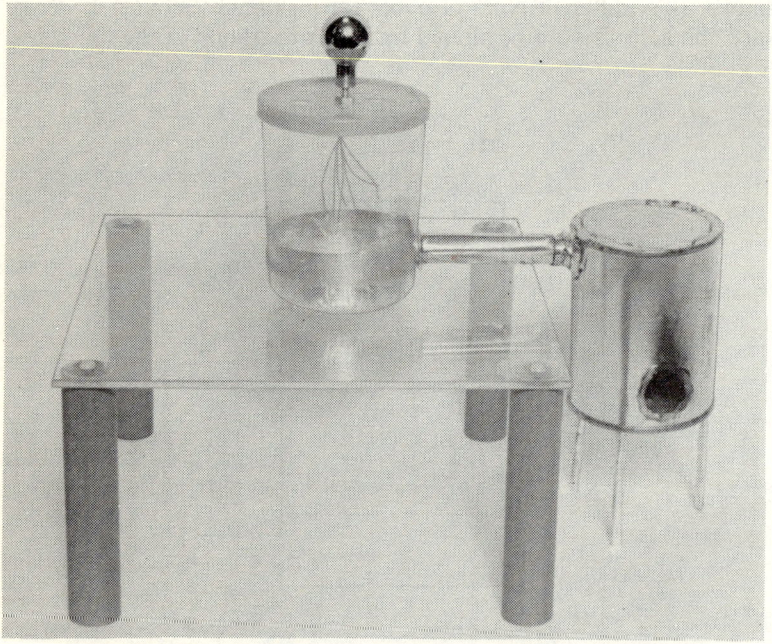

Fig. 7-9. The smoke removal precipitator project assembled with the plastic experiment table.

Stack the plastic sections and the foil sections as shown in Fig. 7-11. Make sure that there are plastic sections over both of the outside foils to prevent electrical shock when the speaker is in use. Cut two cardboard rims, or frames, approximately ¼ inch wide to fit around the edge of the plastic sections and cement the rims in place (one on each side).

Materials List for Electrostatic Speaker

Wooden picture frame (See text.)	Hookup wire
Household plastic wrap	Tube-type audio amplifier (See text.)
Aluminum foil	

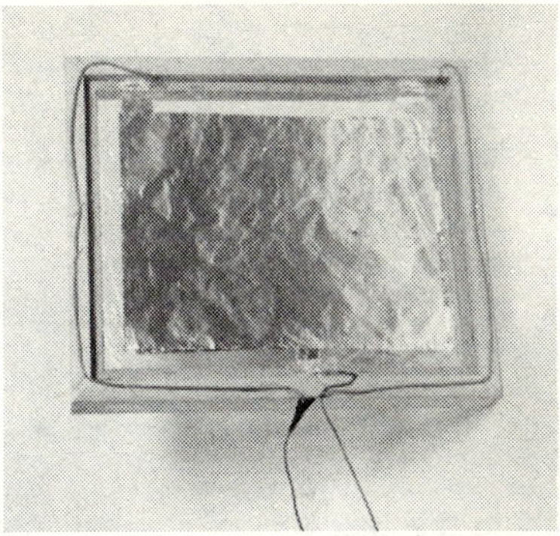

Fig. 7-10. Electrostatic speaker project.

Staple the plastic-foil assembly to the inside of the wooden frame. Make sure that none of the staples go through any of the foil sections, shorting them together. Connect wires to the foil tabs by stapling

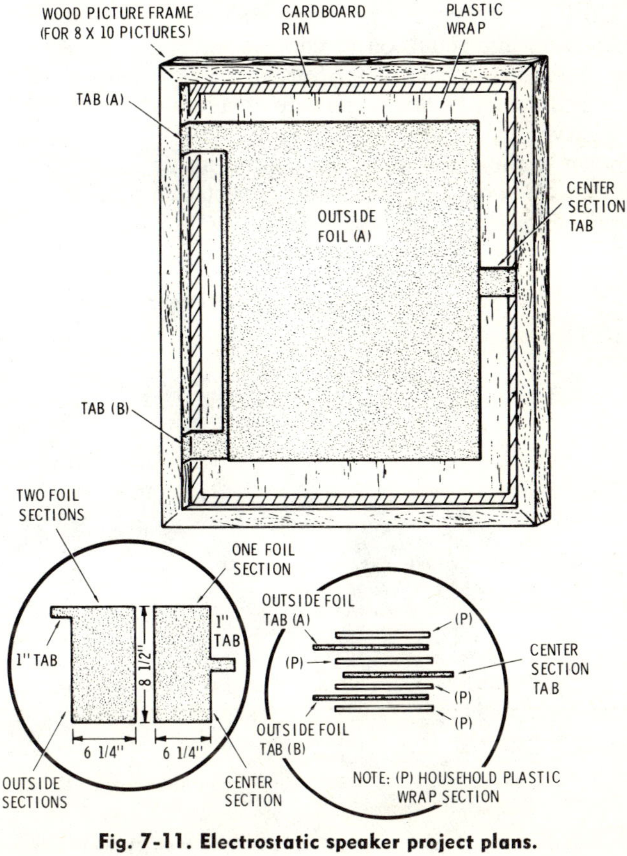

Fig. 7-11. Electrostatic speaker project plans.

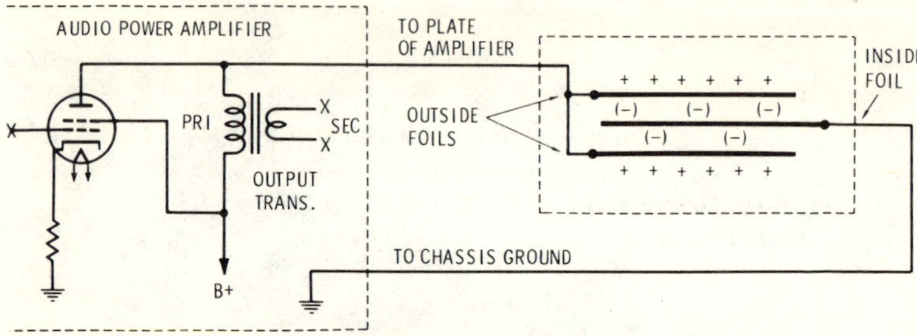

Fig. 7-12. Electrostatic speaker theory.

them together to the wooden frame. Hold the wires in place by additional staples to the wooden frame. The foil tabs and bare wires should be insulated with plastic tape to prevent electrical shock.

Electrostatic Speaker Theory and Operation

The electrostatic speaker is based upon the principle of electrostatic attraction and repulsion of charged materials. As shown in the drawing of Fig. 7-12, the metal foils of the speaker are to be connected to a source of high voltage, with an audio voltage superimposed on the high voltage.

As shown in the schematic diagram, the inside foil is connected to ground, and the two outside foils are connected in parallel to the plate of an audio power-output tube (6V6, 6L6, 6K6, etc.). Thus, the

Fig. 7-13. Electrostatic speaker connected to a tube-type audio power amplifier.

outside foils are dc biased to a high value of B+ voltage (250 volts minimum), and the inside foil is at ground (B−).

The amplified audio voltage across the primary winding of the output transformer (secondary connected to conventional speaker) is in series with the high voltage supplied to the outside foils. As the audio is ac, the polarity and amplitude changes affect the dc voltage potential between the inner and outer foils; the foils become attracted and repelled as the potential changes. Thus, the air is vibrating at an audio rate that can be heard as audible sound waves.

The electrostatic speaker is designed to operate with tube-type audio amplifiers (Fig. 7-13), connected as shown in the schematic, and will not work with transistor amplifiers because of the low dc voltages. The electrostatic speaker requires high voltage to operate (the higher, the better—to about 1000 volts). You can experiment with different sizes of speakers; larger speakers are more efficient for low-frequency notes. The size shown in the project is good only for high-frequency (tweeter) use. The output audio volume will not be high, because of the mechanical limitations of the unit.

APPENDIX

Construction Notes

METAL KNOBS

Many of the construction projects (the electroscopes, Varley machine, electrostatic motors, etc.) call out metal knobs (or metal balls) as terminals. The size of these knobs is not critical. For most of the projects, we used 1-inch-diameter metal, ball-shaped knobs purchased at automobile supply stores; these knobs are actually replacements for control knobs on the automobile dash. Metal, ball-shaped knobs designed for cabinet drawer pulls (purchased at hardware stores) were also used for projects.

The metal knobs need not be ball shaped, but they should have rounded edges. If the metal knobs cannot be purchased in your local stores, you can make satisfactory ones by rolling aluminum foil into balls, by cementing aluminum foil over wood balls, or by painting wood balls with metallic paint. Heavy solid wires, shaped into 1-inch loops, will also work, or you can try thick, metal discs with rounded edges.

PLASTIC

Most types of clear acrylic sheet and tube plastic (from hobby shops or hardware stores) will work. The sheet plastic usually comes covered with adhesive paper. To protect the surface from scratches, it is best to leave the protective paper on the plastic until after it is sawed, drilled, or cut to shape. If long lengths of plastic tubes are required for a project (such as the Kelvin water drop generator), you can usually save some money by buying plastic water supply pipes, made of polyvinyl chloride (PVC), from a building supply store.

Green, Charles,
1891-.

Electrostatics
handbook

Green, Charles
1891-
Electrostatics handbook

duplicate card 12/16/92

Index

A

Amber, 8
Ampere, 10
Amperè, André Marie, 10
Atom, hydrogen, 12
Atoms, 11
Attraction, mutual, 15

B

Bennet, Abraham, 10

C

Charging
  by contact, 15
  by induction, 16
Cloud chamber, 104-108
  construction of, 104-106
  materials for, 105
  theory and operation of, 106-108
Conduction experiment, 19
Conductor, 12
Contact, 17
Cord, hempen, 9
Corona discharge, 13
Coulomb, 10
Current, electron, 12

D

de Cisternay Dufay, Charles Francois, 9
de Coulomb, Charles Augustin, 10
Democratus, 8
Detectors, electrostatic, 53-71
Dielectrics, 14
Discharge, corona, 13
Distribution of electrical charges, 17-18

E

Electret, 109-110
Electric
  current, 12
  fields and charged objects, 14
Electrical charges, distribution of, 17-18
Electrifying
  a plastic comb, 18-19
  by rubbing, 15
Electron sink, 17
Electrons, 11
  free, 12
Electrophorus, 10, 28-30
  construction of, 28-29
  materials list, 29
  theory and operation, 29-30
Electroscope
  gold-leaf, 10, 57-61
  shielded, 61-63
  twin-electrode foil, 63-66
Electrostatic
  detectors, 53-71
  energy, forms of, 7
  excitation, methods of, 17

Electrostatic—cont
generator(s), 11, 23-52
water drop, 11
induction, 15-17
speaker, 117-120
construction of, 117-120
materials for, 118
theory and operation of, 120
voltmeter, 66-71
construction of, 66-69
materials for, 69
operation and experiments, 69-70
theory of, 70-71
Experiments, 18-21
conduction, 18
repulsion of like charges, 21
table, 110-111
water-drop, 20-21

F

Farady, Michael, 10
Forms of electrostatic energy, 7
Four-brush disc motor, 97-100
construction of, 97-99
materials for, 99
theory and operation of, 99-100
Franklin, Benjamin, 9
Franklin motor, 83-87
construction of, 83-85
materials for, 85
operation and experiments, 85-86
theory, 86-87
Franklin's Leyden jar, 79-81
materials for, 81
Free electrons, 12
Friction, 17
generator, rotating, 23-27

G

Galvanometer, mirror astatic, 11
Generator
electrostatic, 11, 23-52
Van de Graaff, 46-52
water-drop, 40-46
Wimshurst electrostatic, 39-40
Gilbert, William, 8
Gilbert's versorium, 8
Glow lamps, 13-15
Gold-leaf electroscope, 10, 57-61
construction of, 57-59
materials list, 59
operation and experiments, 59
theory of, 60-61
Gray, Stephen, 9

H

Hauksbee, Francis, 9
Hempen cord, 9
Henry, Joseph, 10
History of electrostatics, 8-11
Hydrogen atom, 12

I

Indicating devices, pith-ball, 53-55
Induction, 17
electrostatic, 15-17
Influence machine, 10
Insulator, 13, 14
examples of, 14
Ionization, 13
Ions, 13

J

Jar, Leyden, 9

K

Kelvin water-drop generator, 40-46
construction of, 40-44
materials list, 43
operation, 44-45
theory of, 45-46

L

Lamps, glow, 13-15
Law, Ohm's, 10
Leucippus, 8
Leyden jar, 9, 76-79
construction of, 76-77
materials for, 77
theory, 77-79

Lightning rod, 10
Like charges repel, 16

M

Machine
  influence, 10
  Varley, 31-38
Methods of electrostatic induction, 17
Millikan, Robert J., 11
Mirror astatic galvanometer, 11
Molecule, 11
  water, 12
Musschenbroek, Professor Van, 9
Mutual
  attraction, 15
  repulsion, 15

N

Neutrons, 11
Nollet, Abbe, 9

O

Oerstead, Hans Christian, 10
Ohm, Georg Simon, 10
Ohm's law, 10

P

Paper capacitor, 81-82
  construction of, 82
  materials for, 82
Pith-ball indicating devices, 53-55
Plastic comb, electrifying, 18-19
Plastic-disc motor, 87-90
  construction of, 87-88
  materials for, 88
  operation of, 89
  plans for, 88
  theory of, 89-90
Plastic-drum motor, 90-93
  construction of, 90-91
  materials for, 92
  plans for, 92
  theory of, 89-90
Potential, 13
Protons, 11

R

Radiation detection
  early forms of, 101-102
  electroscope test for, 103-104
Repulsion
  mutual, 15
  of like charges experiment, 21
Rod, lightning, 10
Rotating friction generator, 23-27
  adjustment and test, 26-27
  construction of, 23-26
  materials list, 26
  theory of, 27
Rutherford, Lord, 11

S

Shielded electroscope, 61-63
  construction of, 61-62
  materials for, 62
  plans for, 63
Sink, electron, 17
Smoke removal precipitator, 114-117
  construction of, 114-115
  materials for, 115
  theory and operation, 116-117

T

Thales, 8
Thompson, William, 11
Twin-electrode foil electroscope, 63-66
  construction of, 63-65
  materials list, 64
  theory and operation of, 65-66
Two-brush disc motor, 93-97
  construction of, 93-95
  materials for, 95
  theory and operation of, 96-97

U

Unlike charges attract, 16

V

Van de Graaff generator, 46-52
  construction of, 46-50
  materials list for, 49

Van de Graaff generator—cont
operation and experiments, 50-52
theory of, 52
Van de Graaff, Robert J., 11
Vapor trails, types of, 108
Varley, C. F., 11
Varley machine, 31-38
construction of, 31-35
materials list for, 36
operation of, 35-36
theory of operation, 36-38
Versorium, 55-57
construction of, 55-56
materials list, 56
theory and operation, 56-57
Volt, 10
Volta, Allesandro, 10
Voltage, 13
von Guericke's electrostatic generator, 8
von Kleist, E. G., 9

W

Water
bottle Leyden jar, 74-76
materials for, 76
operation and experiments, 74-75
-drop electrostatic generator, 11
-drop experiment, 20-21
molecule, 12
Wimshurst electrostatic generator, 39-40
Wimshurst, James, 11